Moonshadow

The Story of the Total Eclipse

Moonshadow
The Story of the Total Eclipse

Terry Manners

CHAMELEON

André Deutsch Ltd is a subsidiary of **VCI plc**

www.vci.co.uk

Text copyright © **Terry Manners**
Design copyright © **Essential Books**

Design: **Neal Townsend** for **Essential Books**

1 3 5 7 9 10 8 6 4 2

Printed by **Butler & Tanner Ltd, Frome and London**

A catalogue record for this book is available from the British Library.

ISBN 0 233 99680 X

First published in Great Britain in1999 by
Chameleon Books
an imprint of André Deutsch Ltd
76 Dean Street
London W1B 5HA

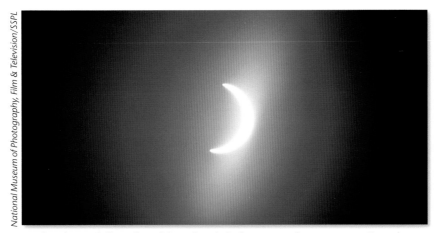

National Museum of Photography, Film & Television/SSPL

The almost fully-eclipsed Sun shortly before second contact over Fatephur.
Photographed through a small reflecting telescope using a mylar filter.

Contents

Introduction

by Steve Bell **HM Nautical Almanac Office**

A total eclipse of the Sun is the greatest natural spectacle most people will ever see. No amount of colourful prose or beautiful photographs can prepare you for the experience. It is one of the very few natural events that can put you in touch with the feelings of wonder and foreboding that our ancestors must have felt in witnessing such an event. Visually, it is awe-inspiring, but the sensory input is not limited to your sight. It is fascinating to observe the effect that the eclipse has on the people around you, to listen to their reactions and to savour the almost tangible atmosphere that the event generates. By way of an introduction, I thought it might be helpful to recall my memories of the 1998 Caribbean total eclipse of the Sun.

For many years I had thought about going to see a total eclipse of the Sun, especially after studying reports of the event over northern India in October 1995 but time and opportunity had never coincided with each other. With some encouragement from my friend and fellow astronomer, Dr Don Pollacco, we made preparations to see the total eclipse of the Sun on 26 February 1998. The island of Antigua was selected as the most accessible place from which to observe it and, coincidentally, to provide an attractive destination for a much-needed holiday as well! My colleague, Margaret Penston, joined our eclipse expedition so that we could quite justifiably claim to be the 'Antiguan outstation' of the Royal Greenwich Observatory. As it turned out,

Previous pages *Bathed in beauty. The stunning landscape of Zanzibar on the morning of 23 October 1976 during a total eclipse. (Michael Maunder)*

The great spectacle begins: First contact of a total eclipse seen from Fatephur Sikri, a ruined Hughai city near Accra in northern India, on the early morning of 24 October 1995. The Moon begins to move in front of the Sun and blocks off its light.

this was the last eclipse expedition by members of the RGO before its closure at the end of October 1998.

After two or three days of scouting around the island for an open, flat site with a good view of the south-western horizon, we selected a cricket pitch in a village called St Philips, in the south-eastern corner of the island. Our last day of scouting, the day before the eclipse, was completely overcast. As a result, we were very concerned that our attempt to see totality was going to be a wash-out. This was my first experience of what can go wrong during an eclipse

The wonder of an eclipse: Second contact over Fatephur. The corona, the Sun's outer atmosphere, can be seen, along with solar flares or prominences which appear as red tongues of light.

10

expedition. However, after giving a talk to guests at the Long Bay Hotel that evening, the weather cleared to reveal a beautiful dark tropical sky punctuated by hundreds of stars. The following day, we arrived at St Philips about three hours before the eclipse was due to start. Armed with single-lens reflex cameras, video cameras and more tripods than you could shake a stick at, we made our preparations for the eclipse. The weather looked good to the south of us but the north-western sky appeared very ominous. One o'clock soon came round and the beginning of the eclipse was upon us.

'First Contact' arrived just as our predictions had indicated. A slowly increasing bite was being taken out of the Sun's bright disk as we watched the Moon make its presence felt. Video cameras sampled the partial phase while cameras clicked at irregular intervals freezing the moment for posterity. You could understand why the ancient Chinese thought that the first half of the eclipse was caused by a dragon slowly consuming the Sun. In the meantime, quite a crowd had gathered on the cricket pitch, made up of both tourists and villagers. Everyone seemed to be well prepared for the event with their aluminized mylar eclipse viewers at the ready.

During the partial phase, I took the opportunity to test some of the specialist solar viewers I had brought out with me. They may not have been the most elegant of devices, but they certainly protected my eyesight. About 25 minutes before totality, one of the local grandmothers put all of the

astronomers to shame by spotting Venus in the western sky. At this point, we heard it was raining in St Johns, Antigua's main town, only 15 km to the north-west. As we looked around us, we found ourselves bathed in a bizarre metallic grey light. As the bright photosphere of the Sun diminished in size, images of the Sun formed by the foliage of the bushes and trees became more crescent-shaped. Shadows cast by objects around us also seemed noticeably sharper. Alas, there was not much wildlife about, so we could not tell if the birds were going off to roost as they would normally do at nightfall. However, we did notice that the goats sharing the cricket pitch with us had disappeared!

By 2.30 p.m. it was getting noticeably darker and the temperature had dropped by several degrees. Only two minutes to totality. The Sun was now a thin sliver of light. People around us were cheering and Handel's *Messiah* could be heard playing on the local radio station. Don shouted for people to watch out for the shadow approaching from the direction of Montserrat, 50 km to the south-west. Some of us also looked around for shadow bands without success. These parallel bands of alternating light and dark are low-contrast features caused by the passage of light from the thin crescent of the Sun through cells in the upper atmosphere. We found out later that guests at the Long Bay Hotel had seen the shadow bands just before the eclipse. They described these elusive features as giving the impression that the white sandy beaches were rippling as the bands travelled across the ground at walking pace.

With only seconds to go, thin wispy cloud formed around the Sun. We could see some Baily's Beads as the last shafts of sunlight passed through valleys on the limb of the Moon. One by one the beads disappeared and through the cloud we could see a beautiful 'diamond ring'. Moments later, the cloud dissipated and 'second contact' had arrived – totality had begun. With Mercury and Jupiter in attendance, the Sun's corona was clearly visible out to a solar diameter or more away. Dotted around the limb were prominences looking like small red dots in the inner corona and the chromosphere. The detailed structure of the corona and its beautiful pearly white colour will stay with me for the rest of my life. People around me simply stood in wonder, savouring this once-in-a-lifetime event.

The south-western sky looked dark, displaying the yellows and oranges of a twilight sky. Behind us, to the north-east, the sky appeared quite bright with puffy white clouds on the blue. The contrast between the two halves of the sky was quite marked. Somehow, I thought it would have got darker than it did. To the south, the island of Guadeloupe stood out clearly against the horizon. Everyone did their best to catch and savour every second of totality. After a little over two and a half minutes, another beautiful 'diamond ring' brought the greatest show on Earth to a close. We all felt that it was the fastest two and a half minutes

we had ever experienced. As the crowd melted away after 'third contact', the dedicated eclipse watchers continued to film the partial phase for the next three quarters of an hour or so until we were clouded out by the cumulus that had affected St Johns earlier on.

On the morning of 11 August 1999, we will have an opportunity to see this awe-inspiring phenomenon here on British soil. The Sun will be close to the peak of its eleven-year activity cycle, so we can look forward to a good coronal display. The eclipse also occurs close to the peak of the Perseid meteor shower, so we might even see a meteor during totality. Sadly, it is very unlikely that any of us will see the next total eclipse visible on the British mainland in just over ninety years time. Let us hope the weather is more favourable than it was for the last total eclipse of the Sun on 29 June 1927, when low cloud spoilt many people's view. Despite the fickle nature of the British weather, it really is an opportunity not to be missed. In fact, I have been told that eclipse watching is every bit as addictive as smoking or gambling. If that is true, then perhaps we will meet in southern Africa on 21 June 2001 for the next total eclipse of the Sun.

Totality: The awe-inspiring moment feared by ancient societies. Here, the eclipse is total over Fatephur. The Moon blocks out the Sun's light, allowing the magnificent, pearly white corona to glow in the darkness.

Prologue

The phenomenon that fascinates the human race

'I caught my breath as I looked up. The grey shadows near me were really men and women but they looked like ghosts. Their faces were dry and dead. This, I thought, is what it must be like to be dead – grey, colourless and quiet, all standing together in an unearthly light, waiting, rather frightened and full of unspeakable revenge.'

These were the words of a *Daily Express* Special Correspondent as he stood in awe on the playing fields of Giggleswick Grammar School, Yorkshire, viewing the total eclipse of the Sun at 6.21 a.m. on 29 June 1927.

Just over seventy-two years later, as we approach the millennium, this natural and spectacular phenomenon will be visible from the shores of Britain once more – on the eleventh minute of the eleventh hour of the eleventh day of August 1999. We will not see another total solar eclipse from our nation for nearly a century. It is the most amazing natural sight you will ever see in your lifetime and one that you will never forget.

In earlier times, scientists and astronomers puzzled over such events. They had worked out mathematically that if an eclipse happened, another one of a similar type would occur eighteen years, ten days and six hours later. This period of time is known as the 'saros'. Mostly, though, eclipses and comets were recorded in early chronicles because they were believed to signal forthcoming disasters such as famine, plague or war. Today, astronomers and scientists mathematically track eclipses back through time to help historians date events in the ancient world. This has led them to discover that historians of long ago were less than scrupulous and would often redate eclipses in order to coincide with great battles, the death of kings and emperors, and great enterprises.

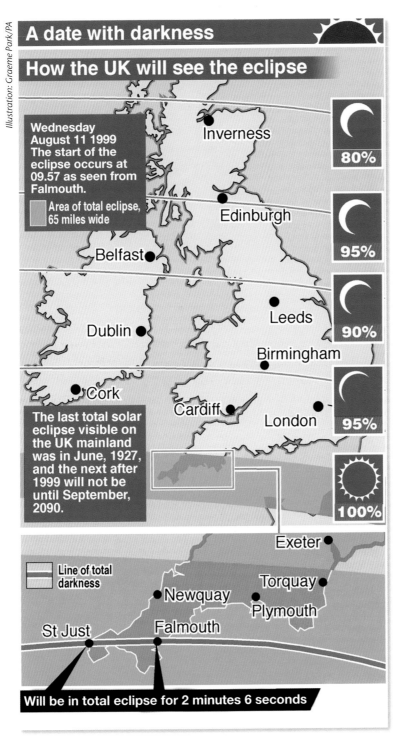

A date with darkness

How the UK will see the eclipse

Wednesday August 11 1999 The start of the eclipse occurs at 09.57 as seen from Falmouth.

Area of total eclipse, 65 miles wide

Inverness **80%**

Edinburgh **95%**

Belfast

Leeds **90%**

Dublin

Birmingham

Cork

The last total solar eclipse visible on the UK mainland was in June, 1927, and the next after 1999 will not be until September, 2090.

Cardiff

London **95%**

100%

Exeter

Line of total darkness

Torquay

Newquay

Plymouth

St Just

Falmouth

Will be in total eclipse for 2 minutes 6 seconds

The total eclipse of August 1999. Graphic shows the path of totality, or line of 100 per cent darkness, over Cornwall and part of Devon. It stretches to Alderney in the Channel Islands, crossing the Cherbourg peninsula, northern France and Luxembourg before moving to southern Germany and on.

Planet Earth

The twilight layering of Earth during a total eclipse, showing the crescent Moon and Jupiter.

Nick Quinn

Natural beauty: Total eclipse over India in October 1995.

Planet Earth

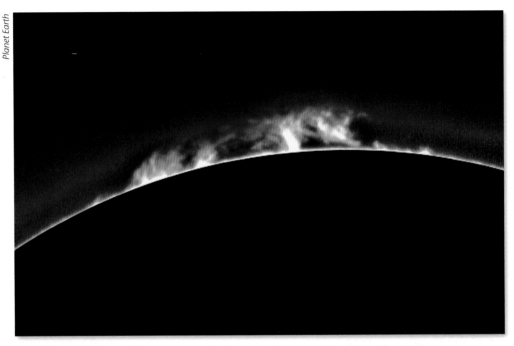

Solar prominences: These spectacular flares, or huge eruptions from the Sun's corona, soar tens of thousands of kilometres into space.

Total eclipses are seen from somewhere on Earth on average every nineteen months, but the position from which someone will see the Moon completely obscure the Sun – called the path of totality – is so narrow that not everyone on the planet will experience it. To understand this shadow, you only have to sit at night and write on a piece of paper with a spotlight shining on your hand. Think of the spotlight as the Sun, your hand as the Moon and your paper the Earth. You will notice that in a certain position the shadow of the hand blocks out the light on the paper and it is almost impossible to see what you are writing. Much of the paper will still be lit, though, because it is outside the path of

totality. The farther you are from the path of totality, the smaller the area of the Sun's surface you see covered – a partial eclipse.

But on 11 August some lucky people in Britain will experience this momentous event. A deepening twilight will cloak a stretch of the British Isles 108 km wide before the daytime sky turns an eerie deep purple and then black. This is the moment that the dark disc of the Moon sits directly between the Sun and the Earth, casting an unearthly pall over the granite rocks and cliff tops of Cornwall and the fields and fauna of Devon.

A deathly silence will engulf the area as seabirds and wildlife hide themselves away,

Nick Quinn

The total eclipse of 3 November 1994, seen from Chile.

many flowers close their petals and millions of people stand in awe. The colours are spectacular; the black disc of the lunar surface is surrounded by the pearly white of the Sun's corona with occasional flame-red prominences. The sky is deep, purple blue and the horizon reddish orange, like a sunset.

This wonder begins at sunrise 400 km south of Halifax, Nova Scotia, and then passes over the Atlantic Ocean towards the British Isles. Its first landfall is the Isles of Scilly, and then the Cornish peninsula close to Land's End. After passing over south-west England and Alderney in the Channel Islands, it crosses the Cherbourg peninsula

and northern France, then the southern tip of Belgium and Luxembourg, before moving on to southern Germany. After passing over Austria, Hungary and the north-eastern tip of Yugoslavia, the phenomenon reaches its maximum duration of totality in the sky above Romania before crossing the north-eastern part of Bulgaria and the Black Sea. It then passes over central Turkey, the north-eastern tip of Syria, north-eastern Iraq, Iran, southern Pakistan and central India, before ending at sunset over the Bay of Bengal, approximately 500 km east of the Indian city of Srikakulam.

The longest period of totality will be seen from Romania and it will last two minutes

Nick Quinn

Breathtaking: The total solar eclipse seen from Morocco.

and twenty-seven seconds. Parts of Cornwall will experience two minutes. A partial eclipse – when the Moon does not completely block out the Sun and casts a shadow called the penumbra – will be seen by people in the north-eastern part of North America, Greenland, Iceland, the rest of the United Kingdom, the Irish Republic, the remainder of Europe, the northern half of Africa, the Middle East and much of Asia as far east as Thailand, and central China, where the myths, legends and magic of an eclipse are still part of folklore.

Nearly two thousand years ago, on 21 June AD 19, there was a similar path of totality to that of 1999. The northern limit skirted the southern coast of England and the southern limit at Jersey. At mid-track totality lasted over four minutes and as it was the summer solstice, the Sun was high in the sky, giving the inhabitants of Guernsey and Alderney the best views and three minutes' totality.

The last total eclipse clearly visible from Britain, seventy two years earlier, took place just after sunrise when the path of totality crossed the coast of Cardigan Bay to the west of Porthmadog, then went on to the east of Colwyn Bay and out over the Irish Sea, crossing the coastline again at Southport and moving north-east of Settle, Richmond, Darlington and West Hartlepool before going out over the North Sea. Totality lasted a little less than 25 seconds and was seen from a strip of land 50 km wide, although most views were obstructed by bad weather. But the interest in the natural phenomenon was so great that alarm clocks sold out all over Yorkshire and Lancashire a week before the event and the Boy Scouts were put on traffic patrol.

Seventy-two years on, millions of people, among them astronomers, physicists and historians will flood into Britain's south-western peninsular to view the eclipse of 1999. And recording this accident of nature will be the solar adventurers, the eclipse-chasers – in other words, a band of professional photographers and scientists who dedicate their lives to capturing this heavenly spectacle on film. Their stories and pictures have made this book possible. They are the lucky ones, going all over the world to experience the magic and might of this solar spectacular. For most of us in Britain, this will be a once-in-a-lifetime experience.

The Royal Astronomical Society is gathering for the eclipse at Fort Albert in Alderney, where the event is particularly significant. The last solar eclipse visible from the Channel Islands was over 1,000 years ago, on 22 December AD 968. It happened shortly after sunrise and lasted only a few seconds across Jersey and Guernsey.

So, how and when do eclipses happen? From the Earth, the Moon and the Sun look the same size, but this illusion is brought about by distance. The Moon, our airless satellite, is just 3,476 km across and only 384,000 km away from us. The Sun is vast: 1,392,000 km in diameter. As it is further away from us than the Moon – 150,000,000 km – it looks smaller.

The Moon circles the Earth every 27 days

and 8 hours, so there are times when it comes between the Sun and us – a solar eclipse. Solar eclipses like the one on 11 August 1999 happen at New Moon, when the Moon is between the Earth and the Sun. The closer the Moon is to the Earth, the bigger it seems and so the more dramatic the eclipse. If it is too far away when an eclipse occurs, it does not appear large enough to cover the Sun completely; this is called an annular eclipse. If it does appear large enough to cover the sun, as it will on 11 August 1999, this is called a total solar eclipse. There are at least two eclipses of the Sun each year, although most are partial. A total eclipse of the Sun can last as long as 7 minutes and 31 seconds, an annular eclipse 12 minutes and 30 seconds.

At full moon, when the Earth is between the Sun and the Moon, lunar eclipses occur. Over time, more lunar eclipses than solar eclipses can be seen from any particular location. Such eclipses do not take place every month because the orbits of the Moon and the Earth are tilted at an angle and most of the time the line-up is not precise enough in space for an eclipse. There are at least two eclipses of the Moon annually, although, if these are penumbral, the Moon is not seen to darken very much. A total lunar eclipse can last 100 minutes or more.

There can be as many as seven eclipses (solar plus lunar) in any one year. In 1935, for example, there were five solar eclipses, four partial and one annular. A total solar eclipse is visible from somewhere on Earth about every nineteen months (that is where our eclipse chasers come in). However, from any one location on Earth, total eclipses take place on average only once in several hundred years. So August in Cornwall and also parts of Devon really is our last chance to see the phenomenon – and we shall be linking ourselves back in time to prehistoric peoples who once stood in awe at this spectacle, just like us.

Recent Total Eclipses

The most recent total solar eclipse seen from our mainland was on 29 June 1927. The path of totality crossed North Wales and northern England, where the weather was mostly cloudy; totality lasted just under twenty-five seconds. A total eclipse was visible from the very north of the Shetland Islands on 30 June 1954.

The Future

Although the Channel Islands will see another total eclipse of the Sun lasting a little over two and a half minutes early on the morning of 3 September 2081, the eclipse of 11 August 1999 will be the very last opportunity to view the event from the UK mainland for another ninety years.

For, on 23 September 2090, the south-western tip of the Irish Republic, south-west England, most of the south coast and the Channel Islands will again see a total eclipse lasting just over two and a half minutes, just before sunset. As the path of totality is over 540 km wide, totality will be seen by much of northern France as well. Most of the UK will witness a partially eclipsed Sun at sunset.

Ground view: Images of the crescent Sun visible in the shadows beneath the trees during the annular eclipse seen from Arizona on 10 May 1994. The phenomenon is caused by gaps in the foliage acting as pinhole cameras which focus the solar image.

The Twenty-Second Century

The first of two total eclipses of the twenty-second century will occur in quick succession, according to Steve Bell. On the morning of 3 June 2133, there will be an eclipse over the Outer Hebrides, the Shetland Islands and the north-western tip of Scotland. Stornoway will experience approximately two and a half minutes of totality, being on the southern side of an eclipse path 280 km wide.

The second eclipse will be shortly after sunrise on 7 October 2135, over central and southern Scotland and north-east England. The path of totality will be approximately 180 km wide and the duration a little over two and a half minutes of viewing from the Glasgow area. Seven years later, the Channel Islands will experience another total eclipse of the Sun early on the morning of 25 May 2142. St Helier, on Jersey, is on the northernmost side of the path, which will be 180 km wide, and will see approximately three minutes of totality.

After nine years, another total eclipse will occur in the early evening of 14 June 2151. It will be seen as total from northern Ireland, north Wales, south-west Scotland, northern England, the Midlands and East Anglia. The path of totality will be 240 km wide; Leeds will see two and a half minutes' totality.

Baily's Beads

These are blobs of light, like beads, seen at

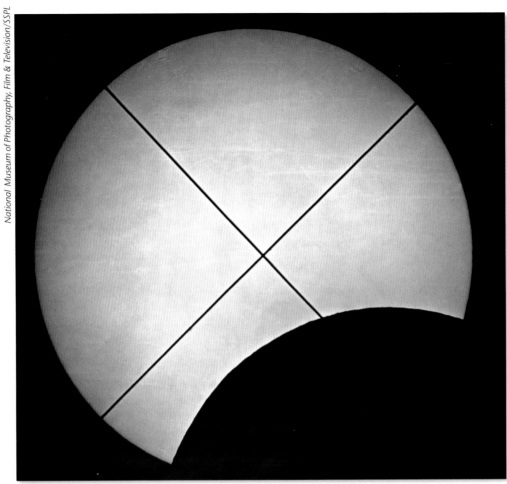

National Museum of Photography, Film & Television/SSPL

One of the earliest pictures of a partial eclipse of the Sun – 16 July 1860.

the beginning and end of the total phase of the eclipse when the thin slice of the Sun that is visible appears to be broken up. They are named after British astronomer Francis Baily (1774–1844), who first suggested the explanation for this phenomenon: that the beads occur because the edge of the Moon is not smooth but jagged with cragged mountain peaks.

The Saros Cycle

Eclipses of the Sun and Moon occur in a pattern that repeats itself every 6,585.32 days (just over eighteen years), a fact that has been known since ancient times. This period is called the saros, a term taken from a Babylonian word and first used by astronomer Edmund Halley. During the saros cycle, the Moon completes 239 orbits of the Earth, returning to a similar position to the one it had occupied eighteen years and ten days before.

The Saros Series

This is a number of saros cycles over a period of time. The 1999 eclipse is the twenty-first in a total of 77 eclipses in the Saros Series numbered 145. Lasting just under 1400 years, the series consists of 34 partial eclipses, 41 total eclipses, 1 annular and 1 annular total eclipse. The series began on 4 January 1639 as a partial eclipse in the North Polar regions and will end on 17 April 3009 in the South Polar area. As Saros Series 145 matures, the duration of totality for each eclipse will then gradually increase to more than seven minutes for the eclipse of 25 June 2522.

Length of Eclipses

The longest possible duration of totality is 7 minutes and 31 seconds and the longest total solar eclipse of the twentieth century was on 2 June 1955 (7 minutes and 8 seconds).

The most recent long eclipse was on 11 July 1991. That came to 6 minutes and 53 seconds and was visible from the Pacific, Central America and Brazil. The next longest is on 22 July 2009. That will be a duration of 6 minutes 39 seconds. Both the 11 July 1991 eclipse and the 22 July 2009 eclipse belong to Saros 136.

Corona

When there is a total eclipse of the Sun, the corona, which stretches millions of kilometres into space, can be seen as a pearly white halo around the black disc of the Moon. At other times, the Sun is too bright for the faint corona to be visible, although it extends for many times the size of the Sun's yellow disc. The light of the corona comes partly from glowing atoms and partly from sunlight scattering off electrons and dust particles, which soar away from the Sun and are replaced every second by more particles rising from its turbulent surface. As they spread out invisibly through the solar system, they create solar winds, rushing by the planets into deep space, where they join other particles speeding into the depths of darkness from other stars. Sometimes they are trapped by the magnetic pull of our planets, giving rise to the displays of aurora in the polar regions of the Earth.

Chromosphere

This is a sphere of colour: a thin layer of red, glowing gas, mainly hydrogen, around the Sun, briefly visible at the start and end of totality. Sometimes it is pink: the colour that hot hydrogen gas glows. Prominences are sometimes seen as pink flames, tongues of hot gas leaping up from the Sun's surface. In fact, it was during the eclipse of 1851 that the chromosphere was named the 'Sierra' by astronomer George Airy because he believed that its jagged upper edge was created by solar mountains. It was only when experts realized, much later, that the Sun was a ball of gases that the term 'chromosphere' was introduced.

A Lunar Eclipse

At full moon, the Moon and the Sun are on opposite sides of the Earth. Sometimes all three bodies line up in the sky with the Earth exactly between the Sun and the Moon. This blocks the light of the Sun and stops it reaching the Moon. The Earth's shadow then darkens the lunar surface. Any of us can see an eclipse of the Moon as long as the Moon is above the horizon and only when the Moon is full. The Earth's shadow has two parts – the umbra and the penumbra. The umbra is the inner part, which is dark. The penumbra, the outer greyish part, gets lighter towards its outer edge. If the Moon only goes through the penumbra, the dimming can hardly be seen by human eyes. In a lunar eclipse, the Moon is not normally completely dark because some sunlight is scattered towards the Moon by Earth's atmosphere. Usually, the Moon appears a deep, coppery-orange colour, even during totality. However, the colour and brightness are variable from one eclipse to another. They depend on factors such as the amount of volcanic dust and cloud in the atmosphere at the time.

Annular Eclipses

The distances between the Earth and the Sun and Moon vary slightly and that means their apparent sizes vary sometimes too. Occasionally, an eclipse occurs that would be total except that the Moon looks too small to cover the whole Sun. The result is an annular eclipse, where a bright ring of the Sun stays visible. Annular means ring-shaped. The corona cannot be seen during an annular eclipse because even a thin ring of the Sun makes the sky too bright. In some cases the corona can be seen, but only when the ring is very, very narrow.

Partial Eclipses

On either side of the path of totality, there are areas where the total solar eclipse can be seen as partial. The farther you are from the path of totality, the smaller the area of the Sun's surface you see covered.

Prominences

These are surges of glowing gas rising from the solar surface. The largest appear as huge arches, sometimes as high as ten Earths piled on top of each other. They can last for several hours before collapsing back. Prominences follow lines of magnetic force and seem pinkish when seen at the edge of the Sun during an eclipse. Some, however, are so violent and explosive that they shoot off the Sun's surface at speeds as great as 1,000 km a second, soaring into space because the Sun's gravity is not powerful enough to pull them back.

Planet Earth

Wonder through the clouds. A partial annular eclipse of the Sun seen from Whangarei, New Zealand, on 16 January 1991.

Chapter One

Our Solar System

In the beginning was there really a beginning?

Nothing lasts forever, not even the universe. Just as it was born, it will surely die. Our Sun, our Moon, our planets and other galaxies have their own life span just as humans do. And humans could be here on this whirling ball of iron and rock we call Earth just by chance. Our tiny world was spinning around the star we call the Sun in our solar system for nearly a billion years before conditions were right for the creation of life.

The universe is many times older than that – 15 billion years. It is everything that exists: the planets, stars, moons, galaxies and space. And it is vast: between 90 and 99 per cent of its matter is dark, unseen and unknown. What we can see through telescopes on Earth makes up only 1 per cent of the material of the cosmos. Some astronomers see the universe as a living thing and the Earth as just a molecule within it. They say we are like microbes trying to work out the structure of the human body

we are in, battling to come to terms with the fact that the giant entity around us is a living being.

The human race has only existed on Earth for about 750,000 years; humans who cultivate their own food and form organized community life for only 8,000 years – a speck of dust in time. Most of our discoveries have happened only over the last 500 years, and we still really know so little. We cannot even properly explain how the first living cells appeared on Earth 3.5 billion years ago. All we know is that without water, our plant life could not have flourished and without plant life, oxygen would not have been released into the early atmosphere to make air for animals to breathe.

And who is to say that the universe needs humans or that there are no non-human scientists out there hundreds of millions of light years away who will never be able to reach us? All we know has come from

The Earth, our home, a whirling ball of iron and rock. It spun around the Sun for nearly a billion years before conditions were right for life.

probes, satellites and radio telescopes, revealing a tiny fragment of the mystery of quasars, pulsars and black holes. The truth is that our best knowledge comes from an object in space only 384,000 km away – the Moon, used by astronomers, scientists and mathematicians to plot our galaxy.

Our own solar system is the group of planets, moons and space debris orbiting the mass of our Sun under its gravitational pull. This same magnetic pull holds us down on the surface of the Earth and gives us weight. The more mass we have, the more weight; and more weight makes us slower.

The Earth and Moon taken from the spacecraft Galileo.

Venus is terrestrial and has no moons.

Pluto's diameter is smaller than our Moon.

Pluto with its own moon, Charon. They are so near each other, just 20,000 km apart, that they often produce eclipses of our distant Sun as they orbit.

Who will win a 100-metre sprint: an athlete weighing a 100 kg or an athlete weighing 60 kg? The 60-kg athlete, of course, no matter how fit the other one is, because the force of gravity will not drag him or her back as much. This simple fact is the key to the structure of the universe.

We lie in a vast galaxy of 100 billion suns ... and beyond that are other galaxies, up to 100 million light years away. Such galaxies are the building blocks of the universe. The four planets closest to our Sun, Mercury, Venus, Earth and Mars, are 'terrestrial' from the Latin word *terra*, meaning land, because they are small, dense and have hard surfaces. Beyond Mars is a ring of debris called the asteroid belt, and then the planets Jupiter, Saturn, Uranus and Neptune – all largely made up of gases and named 'Jovian' after the Roman god, Jove. The furthest and smallest planet from the Sun is Pluto, 59,000 million km away. It was only discovered in 1930 and very little is known about it as it has never been visited by a space probe. With a diameter of only 2,300 km, it is smaller than our Moon. Pluto sits on the edge of the outer asteroid belt with its own moon, Charon, about half as large as itself. They are so near each other – just 20,000 km – that they often produce eclipses of our distant Sun as they orbit. Astronomers call them a twin planet because through telescopes they look almost joined.

Then come the comets – 100 billion huge frozen icebergs hurtling through space. Astronomers believe they are cosmic

Planet Earth

> It was that fatal and perfidious bark
> Built in th'eclipse, and rigged with curses dark,
> That sunk so low that sacred head of thine
> *Lycidas*
>
> … as when the Sun new risen
> Looks through the horizontal misty air
> Shorn of his beams, or from behind the moon
> In dim eclipse disastrous twilight sheds
> On half the nations, and with fear of change
> Perplexes monarchs
> *Paradise Lost*
>
> **John Milton**

rubble left over from the time the solar system was formed and are in orbit about 40,000 times further from the Sun than the Earth. They are rarely seen, but sometimes, when a comet's orbit is disturbed, it falls past the Jovian planets into our inner solar system, only to fall victim to the heat of the Sun, which evaporates the ice to create a streaming tail that glows in the sky.

The Earth is a spherical ball of rock with a diameter of just over 12,000 km and a circumference of over 38,000 km. The atmosphere on which we depend for breath and existence is about as thick compared with the size of the Earth as the skin of a peach is compared with the size of the peach. The lowest layer of our atmosphere, up to about 14 km above our heads, is called the troposphere and this is where we see weather conditions.

The Earth moves round the Sun once a year and the Moon moves round the Earth once a month. The Moon, the ruler of our night sky, is our satellite, reflecting the Sun and bringing us light. It is 3,746 km across, a quarter as wide as the Earth or as large as

---

Previous page: *Comet Hale-Bopp over Corfe Castle on 9 April 1997.*

NASA

The debris of deep space taken by the Hubble Telescope.

North and South America put together. As it is only 384,000 km away – a minuscule distance in the universe – it is close enough for its gravitational pull to drag the sea upwards and cause the tides.

As well as moving around the Sun, the Earth revolves, taking twenty-four hours to turn around once. As it rotates, one side faces the Sun for some of those hours and it is daytime. Then that side turns away from the Sun, and the Moon can glow down on it. Meanwhile, it is daytime on the other side.

The Sun is more than a hundred times bigger than the Earth 150 million km away – almost 400 times further than the Moon. It is a huge ball of hot gas, mostly hydrogen that

Planet Earth

Egg nebula: Death of an ordinary star taken
by Hubble.

NASA

A young binary, or double, star in Taurus.
They revolve around a common centre of
gravity in different orbits.

will turn into helium, and is 6,000 °C on the surface and 15 million °C at its core, which is about the size of Jupiter. All the stars we see in the night sky are just like our Sun, burning hydrogen into helium at their cores. Our Sun's gravitational pull makes it the centre of our solar system, but it will die one day, cooling into a white dwarf star, its mass squeezed into a solid lump about the same size as our planet. Our world will then cease to exist.

But before the end we will continue to seek the beginning – if there ever was a beginning ...

The Big Bang

Today most astronomers believe in the Big Bang theory. Nothing existed until suddenly, in a million billionth of a second, enough energy to make all the materials in space came together in one huge explosion. This theory is largely based on the beliefs of scientist Albert Einstein, who told the world that energy could be turned into matter and matter could be turned into energy. There is no definite evidence for it yet, but the galaxies in the universe do seem to be racing apart as if from a vast explosion, and space probes have detected faint heat waves left over from a huge blast. So we can accept the idea that time began with a Big Bang, but what caused such a vast explosion? Nothing exists without time, so there was nothing.

Planet Earth

ALH84001,0

Meteorite ALH 84001 from Mars discovered in the Antarctic.

Just after the Big Bang, the universe was unimaginably small, but it quickly expanded in every direction. From around 10 billion °C, it cooled to 3,000 °C, and atoms, the minute units of matter, could begin to form. These atoms were made mainly of hydrogen, which is the simplest and most plentiful substance found in the universe; the rest were more complex helium forms.

Nothing actually stands still in space. The universe is expanding all the time, racing through the blackness after the Big Bang. As the Earth circles the Sun and the Moon circles the Earth, the Sun speeds past other stars. At the same time our Galaxy spins like a wheel and all the other galaxies speed on as well. New stars are created in the gaps left by this never-ending expansion, and so on. The laws of nature apply to the universe as well as Earth – death gives way to new birth.

The Galaxy in which our solar system lives is known as the Milky Way and is made

Planet Earth

The Andromeda galaxy. Its light takes 2 million years to reach us.

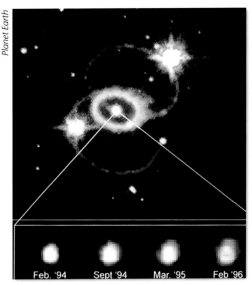

Planet Earth

Feb. '94   Sept '94   Mar. '95   Feb '96

The blast from an exploding star, taken by the Hubble Space Telescope.

up of over 100 billion stars, many similar to our Sun. It sits in space like a flattened disk and is about 100,000 light years across. (One light year is the distance travelled by a beam of light in one year: 9.45 million million km, or 300 million metres per second.) Our Sun and its planets lie about 30,000 light years from the centre.

Between the stars are nebulae, huge dark clouds of powder-sized solid particles and hydrogen gas. These nebulae give birth to suns and planets. Nebulae are so far away they look small, even through powerful telescopes. But in fact each nebula is millions of times bigger than our solar system and contains enough gas and dust to make hundreds of suns. This happens when a supernova, a blast from an exploding star, shakes one of these giant space clouds. The

cloud starts to break up into a group of smaller clouds, each one shrinking inwards from the pull of its own gravity and spinning into matter. Each baby star spins on its axis, glowing dim and red as the force of its inward collapse heats its centre. When the temperature inside reaches millions of degrees, nuclear reactions begin and a wave of energy is blasted outwards. This stops the star shrinking any more and hurls the remains of the nebula into the void. Thus a young star is born. From black cloud to shining star takes about 100,000 years.

Our own young Sun threw out a ring of gas and dust, mostly made up of hydrogen and nitrogen mixed with tiny specks of solid material such as carbon and iron – elements of our world as we know it. This ring began to pick up space fragments and solid grains

The Men On The Moon

Only 12 men have ever walked on the Moon. They spent 160 hours on the surface covering 97 km on foot and by lunar roving vehicle. These astronauts collected 2,196 samples of rock weighing 380 kg, varying in age between 3,100 million and 4,700 million years old. The samples were cut into 39,000 pieces and sent to scientists in 19 countries. More than 30,000 pictures were taken of the Moon and 60 major experiments carried out on the surface. Another 30 were undertaken in lunar orbit

Neil Alden Armstrong

- **born:** 5 August 1930, Wapakoneta, Ohio
- **education:** BS in aeronautical engineering, Purdue University, 1955
- **spaceflights:** Gemini 8, Apollo 11
- **subsequent career:** Chairman, AIL Systems Inc., defense electronics

Edwin Eugene ('Buzz') Aldrin, Jr.

- **born:** 20 January 1930, Montclair, New Jersey
- **education:** BS, States Military Academy, 1951; ScD, Massachusetts Institute of Technology
- **spaceflights:** Gemini 12, Apollo 11
- **subsequent career:** develops ideas for future space transportation systems

of matter, mostly carbon and ice, which stuck together, growing bigger and bigger until their gravity started to pull in other, smaller debris – turning them into the Earth and the planets of the solar system as we know them today.

During thousands of millions of years, the planets in our solar system pulled against each other until their orbits became almost level and stable. They did not turn into suns because they were not big enough or hot enough to start the nuclear reactions necessary to make stars shine. Nevertheless, they have the same ingredients.

New telescopes in the 1920s revealed that there were millions of other galaxies with suns just like ours scattered through the universe. They appeared only as fuzzy blobs through powerful lenses. One was the Andromeda nebula, whose light takes 2

The Moon showing the dark areas Italian scientist Galileo believed were seas.
In fact, they are huge lava plains.

million years to reach us. As for other galaxies, astronomers agree about how fast they are moving away, but not about how far away they are. The further away a star is, the longer it takes for us to see it. So millions of years from now, or tomorrow, we could see light from another new galaxy. But astronomers have not yet been able to see any planets around other stars like our Sun, although they believe that there must be some. If such planets are, in fact, common throughout the universe, other life forms may be common as well.

The Moon
The Moon is the Earth's only satellite, kept in

orbit around us by our gravity. Next to the Sun, it is the brightest object in our sky, more than 2,000 times as bright as Venus. It is 100 times closer to the Earth than any planet and 400 times closer than the Sun itself. The Moon has no atmosphere, no wind and no weather – just a barren landscape of rock and dust.

The Moon is large enough and close enough for its gravitational pull to drag our seas upwards, causing tides around our shores. Land is too firm to respond noticeably to the pull, but water stretches towards and away from the Moon.

Throughout time, the changing face of the Moon has deeply affected people. The new moon was considered the best time to start an enterprise and the full moon was often feared as a time when evil spirits were free to roam. The word 'lunatic' comes from the Latin name for the Moon, *luna*, because it was believed that the rays of the full moon sent people mad.

Astronomers are divided over the Moon's origin. Some say that the Earth and the Moon were formed at the same time from the minute debris and gas of the early solar system; others that the Moon was a body passing our planet that became trapped by our gravitational pull. A Big-Splash theory has even been suggested: that a body the size of Mars once collided with the Earth, splashing debris into space, some of which eventually formed the Moon. This seems the most probable explanation, because we now know, thanks to the USA's Apollo lunar missions, that the Moon's surface is made

Moonbites

● Between new moon and full moon, the Moon is waxing. From full moon back to new moon, it is waning.

● The Moon sequence is: crescent, first quarter, gibbous moon, full moon, gibbous moon, last quarter, crescent.

● The first map of the Moon was drawn by W. Gilbert in around 1600. Telescopes had not been invented and he drew from the naked eye.

up of similar but not identical material to that found on Earth. We also know that the Moon was formed 4,600 million years ago.

Because the Moon has no atmosphere, there has been hardly any erosion on its surface, unlike the Earth's, which has been eroded by weather conditions. Only craters are visible, formed over 4,000 million years ago by the impact of countless meteorites. Each is named after a famous astronomer or philosopher, such as Nicolas Copernicus, the Polish canon who published a book in 1543

Moon Facts

- **Length of day:** 27.3 Earth hours
- **Length of month:** 29.5 Earth days
- **Length of year:** 1 Earth year
- **Average distance from Earth:** 384,000 km
- **Least distance:** 356,000 km
- **Greatest distance:** 407,000 km
- **Temperature:** 100 °C (170 °C on the dark side)
- **Diameter:** 3,476 km
- **Atmosphere:** None
- **Gravity:** 6 times less than that of Earth
  (so US astronauts could jump 6 times higher)

which showed that the puzzling movements of the planets could be more easily explained if they moved in circles around the Sun and not the Earth. Other craters include Newton, Ptolemy, Aristarchus, Halley and Einstein. There is even one named after Julius Caesar, not because he was a leading astronomer, but because of his work on calendar reform. A crater is not the size of the object that made it – it is the size of the explosion the object made. That is why some craters on the Moon are large enough to contain several cities the size of London and have mountains higher than Everest rising from their floors.

The far side of the Moon, or dark side, is covered with mountain ranges, but these mountains were not caused by volcanic eruptions: they are the surviving walls of huge craters. The side of the Moon turned to the Earth has huge lava plains or seas called 'maria' (plural of the Latin for sea, *mare*), caused by the pull of the Earth's gravity, which strained and weakened the lunar crust until molten lava flooded out through cracks. As he studied the lunar surface through the first telescope, Galileo Galilei, the sixteenth-century Italian scientist, believed these to be seas. Today, in his memory, they have names such as the Sea of Tranquillity, Ocean of Storms, Sea of Rain and Sea of Clouds.

Throughout time, the Moon's shadow and bright spots have played games with our imagination. When humans could see the Moon only with the naked eye, they believed the shadows might be a person, a face in the sky looking down and the Man in the Moon became the theme of stories, songs and poems. Some early fiction writers believed that the Moon was a world like Earth with Moonmen watching us.

In 1609, when Galileo first built his telescope, he discovered the mountain ranges and craters. Some of the craters had bright streaks coming out around them and the shadows on the Moon that created the face of a man were the flat, dark areas of the plains he thought were seas.

Moonlight, as we call it, is sunlight shining on the Moon's surface. The Moon has no light of its own. As it moves around the Earth, different parts of it are lit up by the Sun. When the lunar surface is on the opposite side of the Earth from the Sun, it is all lit. We call this the full moon. And when it is on the side of the Earth that is near the Sun, the lit side is away from us and we do not see the lunar surface. In between, as the Moon continues its journey around our planet, it is partly lit. These different shapes that we see in the sky are known as the Moon's 'phases'. Our lunar friend goes from full moon back to full moon in about a month. In ancient times, people used the Moon as a calendar in order to keep a check on dates.

Other Moons

There are sixty-three moons orbiting the planets in our solar system; only Mercury and Venus have none. Nearly all these moons have craters and show a strange variety of surfaces – icy, rocky or covered with brightly coloured chemicals. They nearly all appear to be dead and unchanging, just like our own lunar surface. When astronauts venture further into our solar system, they may well choose to use them as landing places to avoid the risk of being sucked down into destruction on the actual planets.

But one moon, Jupiter's Io, is covered with patches of sulphur and erupting volcanoes. And Saturn's Titan has an atmosphere of nitrogen gas. Twenty-four moons orbit Saturn and three of Jupiter's four largest satellites are bigger than our Moon. One of them is even larger than Mercury. Mars has two moons, named Phobos and Deimos, much smaller than our own. In fact, it is widely believed that they

Sunbites

- The nearest other star to our Sun is Proxima Centauri. Travelling at 16,000 km an hour, one would take 270,000 years to get there.
- The biggest sunspot ever recorded is the Great Sunspot of 1947, discovered on 8 April that year. Its area was 18,000 million km^2.

Planet Earth

All the colours of sunlight are contained within a rainbow.

are captured asteroids. In 50 million years' time, it is possible that tidal forces will disrupt them completely.

The Sun

The Sun was born 4.6 billion years ago and is the centre of our solar system, our local star that brings us day and night, warmth and life. It has no surface, just layers of gases of different densities, mostly hydrogen, which is so compressed at its core that it fuses into helium – the principle of the atomic bomb, now harnessed by us.

This giant ball of fire in the sky loses about 7 million tonnes of material every second, but all the material lost since the Sun

Sun Facts

- **Distance from the Earth:** 149,597,870 km
- **Distance from the centre of the galaxy:** 30,000 light years
- **Temperature:** at the surface, 5,800 °C; at the core, 15,000,000 °C
- **Time taken for its light to reach the Earth:** 500 seconds, or 8 minutes and 33 seconds
- **Diameter:** 1,390,000 km (109 times that of the Earth)
- **Life expectancy:** another 5 billion years
- **Period of rotation:** 27 days
- **Activity:** transforms hydrogen into helium, losing 4 million tonnes of matter every second, which has no effect

started shining amounts to less than a ten-thousandth of its mass.

The flow of energy from the Sun's core, which has temperatures of around 15 million °C, passes in waves upwards through less dense radioactive layers until it reaches the convective zone, where it then becomes turbulent and produces a magnetic field. Above this layer is the photosphere, about 500 km thick, which is the visible face of the Sun – the bright light that we see in our skies – and the home of sunspots. But what we see is dim compared with the core – if it was opened up, our planet would be turned into a cinder in seconds. Sunspots look dark because they are cooler than the rest of the Sun's surface and downward layers, and a cool surface gives out less light than a bright one. Sunspots can grow quickly, reaching

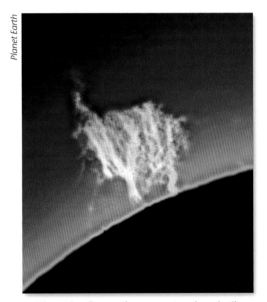

Planet Earth

The solar flares of our constantly volatile Sun. Every minute of the day or night, our very own star converts 300 million tonnes of mass into energy.

Planet Earth

The awesome aurora borealis, caused by atomic particles blasted into space by the Sun. Seen from Northumberland in October 1989.

the size of the Earth in just a few hours. The largest sunspot ever seen was recorded in 1947 and was about 250,000 km long – twenty times the Earth's diameter.

The next layer is the chromosphere, which is hotter and about 4,000 km deep. Then comes an outer layer called the corona, which extends millions of kilometres into space. This is a region of plasma with temperatures around a million °C. During a total eclipse it appears like a crown around the Moon. Both the chromosphere and the corona can only be seen with the naked eye during total eclipses, unless special equipment is used.

The massive energy thrown out by the Sun takes a long time to reach its surface – so long that the heat coming from the surface to us now was created by atomic reactions in its core more than a million years ago.

As the Sun is in a constantly volatile state, converting 300 million tonnes of mass into energy every minute, its layers are always subject to disturbances. Prominences and spectacular solar flares – huge eruptions from the corona that soar tens of thousands of kilometres into space – come from such disturbances in the Sun's magnetic field. Every second the Sun blasts 3 million tonnes of atomic particles into our solar system. This is called the solar wind, and it distorts the magnetic fields of planets such as Earth.

Sunspots, discovered by Galileo, move randomly across the solar surface because the Sun is spinning. But the whole mass of the star does not spin at the same rate because it is not solid. The Sun's equator, for example, takes 25 Earth days to make one complete rotation. Its poles take nearly 36 days to accomplish the same task.

Just as the Sun gives us life, one day it will take life away. By then it is hoped that we will have colonized other worlds. In about 5 billion years from now the massive energy in the Sun's core will make its outer layers, such as the convective zone and photosphere, expand and become hotter. Water on Earth will start to boil and turn into vapour; life forms will suffocate. As the Sun begins to turn into a giant red star, our world will be scorched and humans will no longer exist. As the Sun runs out of nuclear fuel, it will shrink to become a white dwarf star. From the surface of Mars, it will be a dim, white pinprick in the night sky.

The Ozone Layer

Even today we face danger from our local star, for the Sun gives out invisible radiation that destroys living tissue – too much ultra-violet radiation can cause skin cancer and destroy the small plants at the beginning of the food chain. Normally this radiation is filtered before it reaches us by a thin layer of ozone in the stratosphere around our planet. But even though most of the harmful gases known as chlorofluorocarbons, or CFCs, that eat away this layer are now banned from aerosol cans, refrigerators and air-conditioning systems, gases released years ago still have an effect.

But scientists have now discovered a hole in the ozone layer near the South Pole. The layer normally reduces every September above the Antarctic, but this hole in September 1998 was the biggest ever recorded for the time of year. The World Meteorological Organisation reported that the gap in the layer, which protects the Earth from the Sun's damaging ultraviolet rays or radiation, was 15 per cent larger than the previous record and covered a huge swathe of the Pacific and Atlantic Oceans, including the southern tip of South America, an area two-and-a-half times the size of Europe.

Asteroids

The most dangerous asteroid ever recorded is known as 1997 XF11, predicted to sweep past the Earth at 27,000 km per hour at 5.30 p.m. on 26 October in the year 2028. It should miss our planet by 953,000 km – a stone's throw in the scheme of the universe. If it were to hit the Earth, it would have the power of 2 million Hiroshima atomic bombs.

The Earth is showered by meteors about fifteen times a year. These fragments of extremely condensed rock and minerals can vary in size from small boulders to huge jagged slabs. Often they produce beautiful light shows. Particles, mostly smaller than peas, burn up in the atmosphere, causing displays such as the Perseids, which are regularly seen in the August sky.

Three huge meteorites streaked into the Amazon jungle in 1930. They had the power of an H-bomb. Huge tracts of Siberian forest were laid waste by a meteor fireball in 1908. Many scientists believe that 65 million years ago, a giant object from space – 10 km wide – thundered into what is now the Gulf of Mexico, destroying all dominant life on Earth. The Sun was obscured by giant clouds of ash, smoke, dust and smog, bringing acid rain to the planet as volcanoes erupted and earthquakes shook the land. With no Sun to warm the soil or give light and life, a long winter set in. The habitat of the dinosaurs disappeared and the food supply vanished as temperatures plummeted. For between one and ten years, the sky was too dark for vegetation to grow. When the skies cleared and the dormant seeds burst into growth again, there were no large carnivores or grazing herbivores; larger life forms had to begin to evolve all over again.

Today, any threat from meteorites to life as we know it, would probably be met by nuclear rockets like a scene from a sci-fi movie. Even with our limited knowledge of the universe, we are able to track some of the dangerous matter hurtling through our solar system – such as the 2-km XF11. Although we can predict that it will come close to the Earth in the year 2028, only in Halloween 2004, when it passes within 6 million miles of our planet, will scientists will be able work out its true course.

Auroras

Eclipses can be worked out for centuries to come. But the spectacular, awe-inspiring sight of an aurora in our skies cannot. Why? An aurora is caused by the atomic particles blasted into space by the Sun. They soar through the darkness at more than 1,000 km a second. Only one thing can grab them: magnetism – and the metal in the Earth's

Planet Earth

The Northern Lights seen over the Nenana River in Alaska. Solar atoms glow green as they strike the atmosphere after being drawn in by the Earth's magnetic field.

NASA

The planets of our Solar System (from the Sun) ... Mercury, Venus, Earth, Mars, Jupiter, Saturn, Uranus and Neptune. Pluto is further away — 59,000 million km from the Sun.

The Sun and Its Planets

This table shows the distance from the Sun of the planets in our solar system; their orbital period (time spent circling the Sun) and axial rotation (the time it takes to complete a full turn).

Planet	Distance in km	Orbital period	Axial rotation
MERCURY	57.9 million	88 days	58.2 days
VENUS	108.2 million	224.7 days	243 days
EARTH	148.6 million	365.3 days	23 hours, 56 minutes, 4 seconds
MARS	227.9 million	687.0 days	24 hours, 37 minutes, 22 seconds
JUPITER	778 million	11.9 years	9 hours, 50 minutes, 30 seconds
SATURN	1,427 million	29.5 years	10 hours, 30 minutes, 59 seconds
URANUS	2,870 million	83.8 years	17 hours, 14 minutes
NEPTUNE	44,98 million	163.8 years	6 hours, 7 minutes
PLUTO	5,906 million	248.2 years	6 days, 9 hours, 17 seconds

core is a huge magnet. Its power reaches out into outer space, forming a magnetic field around our planet.

When the atomic particles come near the Earth, this magnetic field draws them into our atmosphere at the North and South Poles like iron filings to a magnet. Just over 150 km above the Earth's surface, these particles strike the atoms of oxygen and nitrogen that give us life. In the battle, the atoms glow green or red, and sometimes other colours as well. People living in lands far to the north and south of our planet's equator, where the magnetism is greatest, witness the awesome sight of a night sky lit up with glowing colours. Some aurorae are a faint, greenish glow, but others have the appearance of a curtain of light, shimmering in front of the stars. Sometimes an aurora can be so strong that it sweeps brilliant, coloured rays like searchlights across our night skies.

Chapter Two

Myths and Legends

The hare, the toad, the dragon and eclipses

The awe with which humankind regards the two glowing orbs that illuminate our sky and nourish life on Earth transcends divisions of culture, creed, place and time. Our oldest artefacts – prehistoric cave paintings and stone monuments – reveal a fascination with the Sun and the Moon, and a belief that they rule our lives on Earth. They do, but not in the way the ancients thought.

The mythology and ideology of virtually all ancient societies feature the Sun, Moon and stars as gods and superstitions about the sky's contents are rife throughout history. Early astronomy eventually found a sister, astrology, and today people still live their lives through messages, signs and symbols they see in the stars.

However, as we begin to approach a new millennium, we scientifically filter our view of the celestial ceiling through telescopes and explain any unusual occurrences with the aid of computer print-outs. Scientists and historians would dearly love to be able to view records of eclipses, comets and the night skies from the first three centuries of the Christian era when the Roman Empire was at its height. But, sadly, such works, often written on papyrus, seldom survived, leaving this period of history in a dark age of natural phenomena.

For the ancients, awe was mingled with fear during total solar eclipses. Few natural phenomena inspired more dread. In civilizations all over the globe, people would beg the gods for forgiveness so that sunlight would return to the Earth. Elaborate rituals would be performed to drive evil spirits away and whole populations would fearfully wait to see what calamitous events, such as floods, assassinations or famine, would follow. Even before the coming of Christ, history recorded eclipses that coincided with ill fortunes, such as defeat in battle or the death of kings. However, some scientists

A shepherd gazing at the stars and seeking portents. As astronomy gained popularity, many people still lived their lives through messages, signs and symbols seen in the stars.

later found such coincidences slightly amusing and believed that certain historians of the time tampered with the truth.

In some cultures, however, these signs were taken so seriously that they helped change the course of history. Art, literature and poetry all pay homage to the eclipse, and the Bible also contains references to the awesome spectacle. Even after the wheels of science and progress had turned sufficiently for men and women to comprehend the scientific reason for eclipses and predict their

The Bridgeman Art Library

The ancient Chinese studied and recorded eclipses for centuries but many believed that the dragon was eating away at the Sun. This seventeenth-century colour on silk depicts the Han Dynasty, 202 BC– AD 220.

coming, it only served to diminish the apprehension with which they were regarded, but not the awe, in which they are still held today.

Before the appearance of the first written word, there were the annals of ancient cave paintings, featuring the Sun, the Moon and the Earth, etched into the walls of prehistory 30,000 years ago. These scanty remnants of early man's existence tell the story of day-to-day life, such as hunting and keeping warm. But some of the drawings offer a glimpse of the sky as well. The Sun, Moon and stars all periodically appear on these primeval canvasses, along with marks resembling comets and meteors, and, so some say, flying saucers. Although they leave many questions unanswered, these cave paintings tell us that, since the dawn of humanity, the thoughts of humankind have been directed skywards.

Just as mysterious as these prehistoric drawings are the massive stone structures which began to dot the Western European countryside several thousands of years later. Like the cave murals, they are silent monuments to the past which leave much unexplained. Yet these megaliths have at least one message for us: their creators' dedication to the sky. Although not all of these primitive stone enclaves relate to astronomy, many undoubtedly do. The most famous is Stonehenge on Wiltshire's Salisbury Plain. For centuries, the large circle of rocks has been the subject of controversy over just about every aspect of its creation and purpose – a debate that continues to

this day. Experts still do not know who built it, or how or why it was built in the first place. But they have established that it was erected in three phases spanning the years 2800 BC–1600 BC.

Throughout all of these phases and up to today, the rising Sun at the summer solstice shines almost directly over Stonehenge's tall, outlying 'Heel Stone' when viewed from the centre of the circle. Astronomer Gerald Hawkins maintains that the monument worked like a computer to predict eclipses. But few scientists believe prehistoric man could forecast the event – the gulf between observing and predicting heavenly movements is immense and was particularly so at that time.

The rest of the British Isles also boasts megalithic sites. Ireland is proudest of Newgrange in County Meath, which is the underground tomb dating back to the fourth millennium BC. The large stone burial chamber is set under a mound of mud and grass. There is only one entrance, which features two rectangular boxed openings. At sunrise in midwinter, light shines directly through it and follows a 19-metre passage before illuminating the tomb at the end. These rays were not meant to be observed by human eyes, but by the dead. Less dramatic sites can be found in Scotland, France and Germany.

Many experts believe these places served as shrines and centres for rituals. If so, they served both an astronomical and theological function. At the start of civilization, the two were nearly always wrapped up together – a pattern that proved difficult to change throughout the centuries. Early scientists and mathematicians would often be persecuted or even put to death for questioning religious doctrine. Virtually all ancient societies who believed in more than one god, held celestial deities in high esteem and virtually all of them interpreted heavenly phenomena as signs from these gods indicating their displeasure with humans.

As the world's first astronomers explained that eclipses were actually caused by the motions of the Moon, mythology maintained that the Sun was being eaten by a monster. Educated thinkers told people not to fear eclipses, but the uneducated masses continued to perform ceremonies to bring back the Sun. Science and religion, it appears, developed side by side.

And so, it is not surprising, then, that the Sumerian civilization, which introduced the written word, cultivated maths and gave us astronomy, also built a massive temple to the moon god, Nanna. His shrine was located in the Sumerian city of Ur, situated along the fertile banks of the Euphrates River (now Iraq). For 500 years before and after 2500 BC, Ur served as the capital and because the city's patron was the moon god (each city had its own), he became the supreme Sumerian deity. According to legend, Nanna (also called Sin) was himself the offspring of two other gods, the air god, Enlil and the grain goddess, Ninlil. Their affair was a sordid one. Enlil raped Ninlil, for which he was banished to the underworld. The goddess followed him, only to give birth to

Nanna while still there. He would have perished but for help from an underworld spirit. It was his tainted background which formed the root cause of the demon attacks which occasionally afflicted him, resulting in eclipses. The Moon recovered only because Enlil was able to predict the attack and warn the water god, who sent his own son to the rescue.

During lunar eclipses, Sumerian kings would take ritual baths to symbolize purification of the Moon.

The explanation of a hungry monster causing eclipses satisfied the civilizations that took over Sumer, notably the Babylonians in 2000 BC. And it made sense to people around the world, as populations from Asia and Europe to the Americas adopted the same theory.

Like their Sumerian predecessors, the Babylonians followed a lunar calendar and it was the priests' duty to add months in order to correct any shortcomings. But their interest in astronomy, and in the Moon particularly, went beyond this calendar and they began to chart our solar system. Their first record of a dated eclipse goes back to 25 July 2095 BC observed in Ur, although some scholars dispute the accuracy of this. They maintain that the oldest absolutely accurate and reliable record of a solar eclipse – on 15 June 763 BC – was made in Assyria, another offshoot of the Sumerian civilization, and is mentioned on tablets now in the British Museum. Meanwhile, the earliest certifiably accurate record of a total eclipse is from 709 BC in China. Eclipse predictions began at

The Bridgeman Art Library

This sixteenth-century oil painting by French artist, Antoire Caron shows astronomers of the time studying an eclipse of the Sun. It hangs in the Paul Getty Museum, Malibu, California.

Images Colour Library

The great passageway into the cairn at Newgrange, circa 3,000 BC. At sunrise in midwinter, light shines directly through it to illuminate the tomb at the end.

I have always read that the world, both land and water, was spherical, as the authority and researches of Ptolemy and all the others who have written on this subject demonstrate and prove, as do the eclipses of the Moon and other experiments that are made from east to west, and the elevation of the North Star from north to south
Letter to the Sovereigns on the Third Voyage, 18 Oct 1498
Christopher Columbus

roughly the same time, 700 BC, with the Babylonians. They did reasonably well with predicting solar eclipses – to within a couple of hours – and much better with lunar eclipses – to within one hour. This was because it is not necessary in predicting lunar eclipses to take geography into account. In fact, there were still inaccuracies with solar eclipses until the time of Newton, when the laws of motion could be taken into account. Until then, prediction was based on cycles.

By 900 BC, the Babylonians had been able to compile an enormous record of eclipses in a study called *Enuma Anu Enlil*. Although an important scientific document because of its celestial observations, the *Enuma* also had a religious significance. Babylonians believed that the gods sent signs, in the form of eclipses and other natural occurrences, to warn of misfortune. If they noticed the omen, they could perform the appropriate ritual and prevent disaster, so keeping track of likely omens was

very important. The *Enuma* recorded 7,000 of these possible portents, with the Moon receiving more mentions than any other celestial body.

The Sumerians did not exist in isolation. Most of their immediate neighbours were less developed, but a little further off was a civilization clearly the equal of theirs – that of the Egyptians. It was located on the fertile banks of another river – the Nile – and its people also had temples to sky gods and a special devotion to their Moon god, Thoth. Their Sun god, Ra, had a higher status and, in an early attempt at believing in just one deity, the Pharaoh Akhenaten tried unsuccessfully to establish the Sun as the only god in 1365 BC.

The Egyptians' greatest astronomical achievement was a solar-based calendar, corrected to $365^1/_4$ days. Their towering architectural feat, the Great Pyramid (*c.* 2600), also had astronomical significance: it was aligned with what was then the North Pole star. But it was the Egyptian Moon God,

Sunbites

- Even though the Sun is 148 million km from the Earth, its rays are still bright enough to damage our eyes permanently. It should never be viewed directly, or through binoculars or a telescope. Galileo eventually went blind looking at it.
- The coronagraph is an instrument that enables the Sun's corona to be viewed without waiting for a total eclipse. It was developed in 1930 by French astronomer Bernard Lyot.

Thoth, however, who represented writing, numerical sciences and time. He had a canine head and human body. The connection of the Moon to dogs was made by many societies, perhaps because dogs howled at night. Other cultures linked the Moon with the rabbit, frog and snail.

The Egyptians introduced a second common reason for eclipses – heavenly conflict. They believed them to be a side effect of the epic struggle between Seth, god of the desert, and Horus, a sky god whose eyes represented the Sun and Moon. Seth tries to seize the throne of Osiris, his brother and Horus's father, leading to conflict. During their battles, Horus's mother, Isis, tries to help her son by attempting to kill Seth, only to take pity on him after attacking him with a harpoon. Angry at her compassion, Horus decapitates her. Sun god, Ra, then allows Seth to punish Horus for his crime by tearing out his eyes. Moon god, Thoth, eventually finds them and restores his vision, symbolizing the restoration of the waning and waxing Moon and the Sun after an eclipse.

On the other side of the world, another civilization was making similar strides. Around 2,500 BC, the enduring fascination of astral phenomena for the Chinese had begun. They recorded more eclipses than any other civilization, with systematic records from at least the eighth century BC. But even before this date, they took note of heavenly bodies. One of the oldest legends about eclipses dates to the third millennium BC, when a pair of court astronomers angered the emperor by failing to predict an eclipse. One day in 2137 BC, the emperor was shocked to see the Sun darken when he had not been warned by the brothers Ho and Hi. Their punishment was death. According to legend, they are buried underneath a plaque bearing an inscription that reads:

Here lie the bodies of Ho and Hi
Whose fate though sad was visible –

The Bridgeman Art Library

The cosmos according to Babylonian astronomers, 500 BC.

The Bridgeman Art Library

Austrian oil painting of the crucifixion with darkened Sun by Egon Schiele (1890–1918). The Gospel refers to this darkening but it is likely to be associated with weather conditions, rather than an eclipse.

> *Being hanged because they could not spy*
> *Th'eclipse which was invisible.*

Although this story is just legend, in reality Imperial Court astronomers did track, if not predict, eclipses from early on. And since that time, those observations have been coupled with worship.

Records that were inscribed on bones from the second millennium BC mention total eclipses in relation to prophecy. The court paid particular attention to sky omens, because the emperor was seen as the 'Son of Heaven'. Since eclipses were signs from the heavens, they were interpreted as direct messages to the ruler. Sometimes they foretold disaster; sometimes they criticized the ruler himself. The annular eclipse of the Sun on 21 May in AD 30 was interpreted as a demonstration of disapproval for the emperor's refusal to accept the recommendations of high officials.

Eclipses could also predict the death of a ruler. On 4 March 181 BC, state records reveal that, '... the Sun was eclipsed and it became dark in the daytime. The Empress Dowager was upset by it and her heart was ill at ease. Turning to those around her she said, "This is on my account." She later died.'

The earliest record of eclipses use the word *shih*, meaning 'to eat'. Accordingly, the Chinese believed eclipses occurred when a giant dragon devoured the Sun. In order to keep this beast at bay, the court and general population would undertake elaborate rituals. State records from the seventh century BC reveal: 'The Sun was eclipsed; drums were beaten and oxen were sacrificed at the temple.'

Although the Chinese discovered the real cause of eclipses as early as the first century AD, similar ceremonies continued until relatively modern times. A more recent account describes high officials decked out in regalia, lighting candles and incense to mark the rite. After making humble pleas to the heavens, people beat loud gongs and drums and brandished torches to scare the dragon away. Practised for thousands of years, these rituals have naturally always been successful.

The Moon has traditionally held a special place in Chinese heritage, although the Moon goddess of yore, Cheng E (also Heng-O), and the Sun, always ranked below other deities. According to Chinese mythology from the first millennium BC, at one time there were ten suns, all sons of the Lord of the Heavens, Di Jun, who took turns – one per day – to circle the Earth. But they grew bored with the arrangement, so they decided to appear in the sky all at once. When they did, everything on Earth began to wither and die. Di Jun ordered them to leave, but they refused, so he sent Yi, a divine archer, to scare them into submission. When he arrived on Earth, he was so angry at what he found that he ended up slaying all but one Sun. When Di Jun found out, he was furious and sentenced Yi and his wife, Cheng E, to remain down below. Cheng E desperately wanted to return to heaven and an astrologer advised her to flee to safety on the Moon. So she drank an elixir for immortality and floated off. But she had drunk too much – it was meant for them both – and Yi was left behind. Worse was to befall her on arrival, for she turned into a toad, with only a hare to keep her company. The hare and toad are often associated with the Moon because of their monthly birth cycles and their nocturnal behaviour.

India also subscribed to the idea that a monster ate the Sun during an eclipse and included an elixir of immortality in the plot. The monster even had his own name and identity. The story of Rahu, who has a tail like a comet and a head like a dragon, appears in the Vedas, the ancient Hindu scriptures. A long time ago, Rahu became jealous when

The Men On The Moon

Charles Conrad, Jr.

- **born:** 2 June 1930, Philadelphia, Pennsylvania
- **education:** BS in aeronautical engineering, Princeton University, 1953; Naval Test Pilot School, 1958
- **spaceflights:** Gemini 5, Gemini 11, Apollo 12, Skylab 1
- **subsequent career:** Vice President, McDonnell Douglas Corporation

Alan LaVern Bean

- **born:** 15 March 1931, Wheeler, Texas
- **education:** BS in aeronautical engineering, University of Texas at Austin, 1955; Navy Test Pilot School, Patuxent River, Maryland, 1960; School of Aviation Safety, University of Southern California, 1962
- **spaceflights:** Apollo 12, Skylab 2
- **subsequent career:** artist

he discovered the gods drinking the nectar of eternal life. So, disguising himself, he began to drink it as well. The Sun and Moon saw him and summoned the god Vishnu. He quickly unleashed his discus and cut off the beast's head. The body crashed to Earth, creating mountains and islands, but Rahu's head survived and rose to the sky. In an eternal quest for revenge, he eats the Moon every month and the Sun during eclipses. Despite this troublesome degradation, the Sun god, Surya, was widely worshipped in classical India, while Vishnu also evolved out of a Sun god background.

Like the Chinese, the Japanese also believed eclipses to be powerful portents and stopped work on days when they occurred. Sometimes eclipses even resulted in a general amnesty, when prisoners were released from jail and forgiven for their crimes. Japanese mythology concerning sky deities and the cause of eclipses also features family conflict, such as the rivalry between a brother and sister. The god Izanagi fathered three children, among whom he decided to divide the lands of the world. He gave his daughter, Amaterasu, control of the heavens, her brother, the Moon god, the realms of the night, and her other brother, Susano, the oceans. Susano objected to this arrangement and, angered by his defiance, Izanagi banished him. Susano began to quarrel with his sister, ultimately throwing a skinned horse into the sacred weaving hall, over which she was sovereign. Frightened by this action, which in some versions results in her injury and, in others, her assistant's

Aerial view of Stonehenge. Some of the sarsens weigh 30,000 kg each – the outer range, with lintels, has a diameter of about 30 m. Astronomer Gerald Hawkins believes the monument works like a computer to predict eclipses.

death, she retreats to a cave, leaving the entire world dark. An elaborate ploy by the other gods then coaxes her out and finally she returns to the sky. Her brief absence symbolizes death, winter and the temporary triumph of evil, highlighted by the hiding and resurfacing of the Sun during a total solar eclipse.

On the shores of the Mediterranean, *c.* 1000 BC, a civilization arose that would set the standard for all of western thought, that of Ancient Greece. Like the peoples of other ancient societies who believed in more than one god, the Greeks built temples that reflected their interest in the sky. The chief god, Zeus, who ruled from the lofty heights of Mount Olympus, was the father of the twins Apollo, the Sun god, and Artemis, goddess of the Moon. These principle gods also had lesser deities associated with their celestial domains (the god of the Sun was Helios; of the Moon, Selene). Ancient writers held that the shepherd Endymion, Selene's lover, was responsible for eclipses. Having been put to sleep as punishment for attempting to rape the goddess, Hera, or as the price for eternal life and youth granted by Zeus (according to different versions of the story), Endymion catches Selene's eye while he lies asleep on Mount Latmus. She

A Teotihuacan pyramid of the Sun (the third largest in the world) in Mexico. The Aztecs believed in astral gods. When there were no shadows from the Sun, they believed the Sun God was visiting them on Earth.

Images Colour Library

falls in love with him and gives birth to 50 of his children without him waking. Greek poets said an eclipse occurred when Selene visited Endymion.

The people of Greece felt just as strongly about the prophetic powers of eclipses as the Chinese. Whereas eclipses influenced Chinese history by reforming the Emperor's behaviour, they changed the entire course of history for the Greeks. The most significant eclipse occurred at a crucial moment in the Peloponnesian War. Greece was then a collective of semi-independent city states, which were always fighting each other for supremacy. In 413 BC, Athens and Sparta were battling it out. The Athenians blundered by sending a force to conquer Sicily, where things went from bad to worse right from the start. The only way to avoid utter defeat was to make a quick retreat. But just as the troops were about to withdraw on their ships, there was an eclipse. Nicias, the Athenian leader, interpreted this as a sign from the gods to stay put. In fact, the delay gave the Spartans enough time to wipe out the whole fleet. This catastrophe in turn lead to the surrender of Athens itself.

Almost 200 years earlier, an eclipse was the cause of peace between the Greeks and the Persian civilizations. The Medes, allies of

the Persians, had become entangled with the neighbouring kingdom of Lydia. The Greek writer and chronicler, Herodotus (c. 484–425 BC), known as the 'Father of History', described the conflict's rather strange conclusion in his records of the Greek–Persian war, written 150 years later:
'There was war between the Lydians and the Medes five years ... They were still warring with equal success, when it chanced, at an encounter which happened in the sixth year, that during the battle the day was turned to night. Thales of Miletus had foretold this loss of daylight to the Ionians, fixing it within the year in which the change did indeed happen. So when the Lydians and the Medes saw the day turned to night, they ceased from fighting, and both were the more zealous to make peace. Those who reconciled them were Syennesis the Cicilian and Lebnetus the Babylonian ... ' [1]Some scholars doubt whether the eclipse really caused the peaceful resolution of the battle, but through this celestial event they have been able to date the battle to 585 BC. Even more dubious is the notion that Thales predicted it. Thales, who lived c. 625–547 BC, was one of the first great Greek philosophers. Although he did formulate a theory of the composition and structure of the universe, it is highly improbable that he correctly predicted an eclipse. The first person to identify the real cause of eclipses was the Greek scholar, Anaxagoras, who was born in 499 BC, almost a century later. He ended up by being exiled for claiming that the Sun was a fiery body larger than Greece.

That was heresy. Banishment was a relatively light punishment, won directly through his friendship with the powerful Athenian statesman Pericles.

Both men figure in another story, presumably fictional, about an eclipse during one of the Peloponnesian wars when the soldiers became frightened by the darkening Sun. Pericles, who was a student of Anaxagoras, demonstrated the cause of the phenomenon with his cloak, explaining the shadows and dispelling their fears. By the time of Aristotle (384–322 BC), the theory of Anaxagoras had gained enough ground for Aristotle to use it in proving that the Earth was round. He asserted that the curved shadow of the Earth, projected onto the Moon during a lunar eclipse, would only be possible if the Earth were round.

By the first century BC, the glorious Greek civilization had been sucked into the Roman Empire. But the Romans did not produce thinkers like Aristotle or Thales, largely because they did not try to understand how the universe worked. Instead, they wanted to make it work for them. One of their greatest achievements was Julius Caesar's reform of the calendar in 45 BC, which at first was misunderstood by members of Rome's leading mathematical college. They inserted leap years every third year instead of four. The error went on for nearly 50 years but was eventually corrected. However, it must have confused historians and astronomers. Finally, they got it more or less right and the Julian calendar, with a leap year every fourth year, was kept up with

Moonbites

- The brightest crater is Aristarchus, which appears prominently when earth-lit.
- The darkest area is the Floor of Grimaldi.
- The first photograph of the Moon was taken on 23 March 1840 by J.W. Draper using a 12-cm reflector. The image was 2.5 cm across and the exposure time was 20 minutes.

regularity from AD 8 until AD 1582, which helped scholars study the universe.

The Romans borrowed shamelessly from the peoples they conquered and even their religion almost entirely copied the Greeks'. For example, they romanized the names of the gods. Zeus became Jove, Artemis became Diana and Selene, Luna. Only Apollo stayed the same. They also continued to believe eclipses were bad omens from the gods and performed rituals to save the Sun. The great historian, Livy, recounts an incident in 188 BC when a 'darkness' in the daytime caused the city to proclaim a 3-day period of prayer. As late as AD 400, 3 lunar eclipses in one year received this reaction:

'Constant eclipses of the Moon alarmed us and night after night throughout the cities of Italy sounded wailings and the beating of brazen gongs to scare the shadow from off her darkened face. Men would not believe that the Moon had been defrauded of her brother the Sun, forbidden to give light but the interposition of the Earth.'[2]

Just as in China, even though people eventually understood the actual reason for an eclipse, religious ceremonies about the event continued.

The Romans were particularly eager to attach important historic events to eclipses, beginning with the founding of Rome, an event incorrectly attributed to 754 BC. Eclipses did, however, occur close to the assassination of Emperor Nero's mother, Agrippina, (30 April AD 59), whom he had killed after taking power, and to the death of the father of Constantine the Great (27 July AD 306). Eclipses could also curse the beginning of new enterprise, like the one in April of AD 238.

An end to the civil strife was made when the boy, Gordian, was given the consulship. There was an omen, however, that Gordian was not to rule for long, which was this: there occurred an eclipse of the sun so black that men thought it was night and business could not be transacted without the aid of lanterns.

As more societies believing in just one god evolved, pagan temples to celestial deities began to be scrapped. Even in the Bible, the Old Testament prohibited Jewish people from worshipping the Moon. Isaiah

Images Colour Library

The Incans built this temple of Machu Picchu in Peru. Its most important part was its Intihuana or 'solar hitching post' which allowed them to tie the Sun to the Earth.

declares: 'Your new moons and your appointed feasts my soul hateth.' But many scholars have argued that certain rites and festivals recorded in the Bible indicate some enduring devotion to these deities. The Old Testament refers to the 'Queen of the Heavens', while Christianity connects the Virgin Mary with the Moon. Many Catholic depictions of her include astral references, often showing her standing with the glowing orb.

Apart from discarding the old faiths, the Bible maintains many of the traditional attitudes towards eclipses. Some biblical passages are reminiscent of the role of eclipses in signalling historically important events. In the Old Testament, Ezekiel describes the imminent downfall of the Egyptian pharaoh by declaring that the heavens will be covered, the Sun will be veiled with clouds and the Moon will not give light. More often, however, eclipses are tied to the apocalypse. Both the Old and New Testament contain several references to this sign. Joel, describing the coming of the Messiah, notes that 'I will shew wonders in heaven, and on Earth blood, and fire, and whirling smoke. The Sun shall be turned into darkness and the Moon into blood, before the day of the Lord comes, the great, the terrible day.'

Many descriptions of the Moon becoming dark refer to the superior power of God. In the New Testament, John in Revelations 6:12 describes his view of Judgement Day: 'There was a great earthquake. The Sun turned black like sackcloth made of goat hair, the whole Moon turned blood red, and the stars in the sky fell to Earth, as late figs drop from a fig tree when shaken by a strong wind.'

This imagery continued well into the Middle Ages and caused people to worry during eclipses, whether the end of the world had come. On one occasion, a chronicler from Croatia wrote:

AD 1239 on the third day from the beginning of the month of June, a wonderful and terrible eclipse of the Sun occurred, for the entire Sun was obscured, and the whole of the clear sky was in darkness. Also stars appeared in the sky as if during the night, and a certain greater star shone beside the Sun on the western side. And such great fear overtook everyone, that just like madmen they ran about to and fro, shrieking, thinking that the end of the world had come.

Even those who understood that an eclipse would not bring on the end of the world continued to believe that it was a sign of foreboding. The medieval historian, William of Malmesbury, wrote that in England in AD 1140:

'... the solar eclipse was so remarkable that men feared the primeval chaos; but soon learning what it was, they went outside and saw stars around the Sun. It was thought and said by many, not mistakenly, that the King [Stephen] would not continue to reign for a year without loss.'

In fact, within the very same year, King Stephen was imprisoned after his army had been defeated.

A fifteenth-century Spanish tapestry from the Museo de Santa Cruz, Toledo. In the following century, Spaniard Francisco Pizarro was to destroy the Incan temples dedicated to the Sun.

Medieval Christians went one step further, believing that an eclipse was a heavenly comment on human behaviour. At the end of the fifth century, the French bishop and chronicler Gregory of Tours wrote: '... the Sun appeared hideous, so that scarcely a third of it gave light. I believe this occurred on account of crimes and the shedding of innocent blood.'

The pagan Vikings of Northern Europe also believed that an eclipse would signal the end of the world. Old Norse texts, compiled

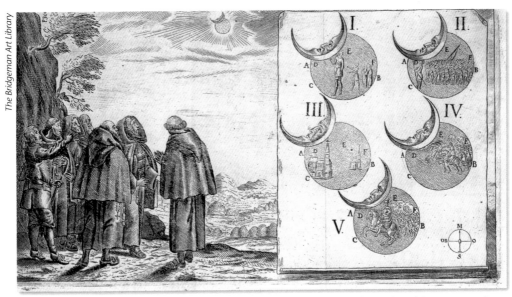

The Bridgeman Art Library

A military attack is plotted in accordance with the phases of an eclipse. (Paul Furst engraving from a seventeenth-century book on the Ottoman campaigns in Europe)

during the twelfth and thirteenth centuries, said eclipses and earthquakes would bring on *Ragnorek*, the 'fate of the gods'. This myth describes how the current order of gods – with the chief god, Odin, at its head – would one day perish in the battle that would end this world and bring on the next generation of man.

Some biblical connections to eclipses, however, merely fulfilled the historical role of helping to date events. Both the birth and death of Christ are linked to an eclipse of some sort and scholars have used this information to work out the approximate dates of his life. Jesus was born during the rule of Herod the Great, then king of the Roman province of Judaea, who lived from 74 to 4 BC. Just before the ruler's death, the ancient Jewish historian, Josephus, notes an eclipse, leading up to the spring festival of Passover. The most probable explanation is the partial eclipse of 12/13 March 4 BC. Assuming that is correct, Jesus was born sometime before March in 4 BC.

The Gospel of St Mark makes references to a darkening of the Sun during the crucifixion but this is almost certainly related to weather rather than eclipses, since Christ's death also occurred near Passover, a holiday celebrated during the full moon. A solar eclipse can only occur during a new moon. A lunar eclipse, however, does occur during a full moon, and on 3 April AD 33, a partial lunar eclipse would have been visible in Israel a little after moonrise. This dating method is not infallible since such dates give Jesus a longer life span than most authorities believe he actually had.

The other great single-god tradition to spring up in the West was Islam in the seventh century AD. Like the Christians, Moslems maintained beliefs about the prophetic nature of eclipses well into this millennium. In 1487, a Moslem writer noted: 'In the month of Safar, the Moon's body was eclipsed and the Earth was darkened. It remained in eclipse for about 50 degrees. People were saying that the demise of the Sultan was coming near. Nothing of what they said happened and the Sultan stayed in power for a long time after that.'

Although the prediction proved to be incorrect, at the time the people expected it to come true. The use of 50 degrees (meaning the amount of space covered by the Sun in the sky during an eclipse) rather than time indicated the superior scientific achievement of the Moslems. They were devoted to the study of astronomy and paid particular attention to celestial observations and mathematics.

The Moslem faith has always had a special relationship with the Moon: the visible crescent, for instance, begins Moslem months and religious festivals. Islam also has traditional rituals related to eclipses: 'The astrologers warned that the Sun would be eclipsed and in Cairo there were callings to the people that they should pray and do good deeds.' Although this account, given by an Arab observer, comes from 1431, the practice of praying during an eclipse continues to this day. In one medieval instance, an eclipse is said to have prevented a caliph – the ruler of the Islamic world – from relocating the pulpit of the prophet Mohammed, the founder of Islam. Interestingly, eclipse prophecy is technically discouraged, as one traditional text reads: 'The Sun was eclipsed on the day that Ibrahim (Mohammed's son) died, and the people said: 'It is eclipsed for the death of Ibrahim.' But the messenger of Allah (Mohammed) said: 'The Sun and Moon are two of the signs of Allah: they are not eclipsed for the life or death of anyone. When you see it, call upon God and pray until it clears.'

Not all Arabic peoples feel so favourably towards the objects in the heavens. In the stifling heat of northern Arabia, a sect of Bedouins curse the Sun each day as she rises because they believe she intends to destroy their lives. They depict her as a bony, cruel old woman who hates and wishes to kill them. An eclipse is a sign of hope, because it means that their suffering could soon end. They cheer on the demon fish woman who pursues her, but, unfortunately, to no avail. She is never successful; after eating the Sun, she invariably vomits her up, unable to stomach the hot, bony body. The Moon is their saviour, a benevolent protector of the community who provides rain and therefore life. Although married to the Sun – it takes him a month to recover from one night together, hence his phases – he is their principle deity and the object of praise and adoration when he appears.

Across the ocean, another culture tells a

similar story of the Sun's pursuer being unable to bear her heat. The Toba of South America say the sun goddess, Akewa, is chased constantly – and occasionally overcome – by a jaguar who then regurgitates her on account of her temperature. The jaguar also chases the Moon, a lazier god who is easier to catch; hence, lunar eclipses happen more frequently. In this culture, the Sun's description relates to her attempt to fight off this jaguar: as Akewa walks across the sky each day, she shrouds herself in clashing weapons – her rays.

Elsewhere among Native American and Inuit populations, the theme of sibling rivalry in the heavens repeats itself. But this time there is a sexual dimension to the traditional hostile relationship. Many legends depict the brother and sister as committing some act of incest. The Greenland Inuits tell of the Sun Goddess Malina and her Moon Brother Annigan. One night the two adolescents were romping around in the dark when Annigan became sexually aroused. Against taboo, he rapes his sister. In her anger she flees to the sky, where he lustfully chases after her for eternity. When he catches her, there is an eclipse. In some traditions, including that of the Alaskan Inuits, the coupling is by consent and only afterwards do the pair realize what they have done. The Panamanian Indians meanwhile describe Sister Sun as so enraged after being raped by her brother Moon that she eternally chases him through the sky.

Other American populations interpreted eclipses as a sign of godly disfavour. The Tewa of Rio Grande believed that Sun Father was retreating because of his displeasure. The Navajo of North America thought eclipses meant great disruption of the cosmos and loss of life. Normal ceremonies were cancelled and people would not eat, staying inside away from the Sun. The Yuchi decorated their bodies and fired arrows at the sky to ward off demons, wailing because the Sun was dying. After an eclipse, the community celebrated, rejoicing because the Sun lived. During lunar eclipses, the Orinoko Indians picked up their farming tools and redoubled their labour, believing the Moon was veiling herself in anger as a result of their laziness.

The Cahuilla Indians of California did the opposite, stopping all work and food-gathering to drive away the invading spirits with singing, dancing and general mayhem. They believed the souls of the dead or those about to die ate the Sun during an eclipse. Warding off these spirits was of the utmost importance – the eclipse meant the Moon was in danger of dying, which would in turn jeopardize the lives of everyone. Since people close to death threatened the Moon, the medicine men predicted who they might be so that they could keep a watch over them to save their lives and their souls.

Great civilizations in South America also regarded the Sun and Moon with reverence, but viewed the Sun as supreme. When the Dark Ages closed in on Europe, the Mayans began to flourish. Located in present-day southern Mexico and Guatemala, they built

*Miniature parchment of a nineteenth-century Persian astronomical
observatory at Merega, showing astrolobes and other measuring instruments.*

Scientists, artists and astronomers were fascinated with the workings of the heavens in the fifteenth and sixteenth centuries. This study, by artist Leonardo da Vinci (1452–1519), shows the outer luminosity of the Moon – lumen cinerum.

an astronomically aligned pyramid and kept a record of celestial events by the fifth century AD. Those tables are known as the 'Dresden Codex', and predicted lunar and solar eclipses. The Mayans also maintained a solar calendar more accurate than Julius Caesar's. Their culture declined around AD 900, swallowed up first by that of the Toltecs and then the Aztecs.

The Aztecs lived in Mexico between 1300 and 1500 and maintained the Mayan belief in the astral gods. They used the

Mayan calendar and another people's urban and ceremonial centre, which they named Teotihuacan. Situated about 48 km north-east of Mexico City, this city was built during the time of the Mayans, but its origins and first inhabitants are a mystery. The Aztecs believed it to be the 'Birthplace of the Gods' (Teotihuacan's literal meaning) and regarded it as the basis for all civilization.

In this giant ruin stand two temples, one slightly bigger than the other. The smaller of them was named the Pyramid of the Moon, the larger the Pyramid of the Sun. The Pyramid of the Sun was aligned with the rising of the constellation Pleiades. By this alignment, it exactly indicated the days when the Sun passed directly overhead. These were very important days due to the fact that there were no shadows at noon and the Aztecs believed that this was the time when the Sun god visited them on Earth.

A different South American civilization – the Incan – would be the last to survive. Originating in Peru, the Incan empire extended from modern-day Ecuador to Chile and included part of Bolivia and Argentina. Like the Mayan and Aztec civilizations, the empire featured massive temples. In the Incan capital of Cuzco stood Corincancha, the temple dedicated to the Sun God, Inti. It was aligned with the winter solstice sunrise, during which the emperor would sit clothed in gold plates so that the Sun could bathe him in its rays. He was the 'Son of the Sun'. The Incas considered themselves to be descendants of the Sun and Moon, so preserving the Sun was of the utmost concern. Ninety-seven km north of Cuzco, and 2,400 metres above sea level, the Incas built the temple of Machu Picchu.

Its most important component was its Intihuana, or 'solar hitching post', so that they could tie the Sun close to the Earth. At the winter solstice, priests performed ceremonies in order to keep the Sun from straying even farther north in its daily path and being lost forever. Signs of the Sun's displeasure were taken very seriously. The Incas believed a solar eclipse was one such sign, an indication of the Sun's anger with the people. Even more disturbing, because the Sun sustained the people, an eclipse could mean the death of the Sun itself, which would end their lives.

As it turned out, the Incas did decline, following Spaniard Francisco Pizarro's arrival in 1532. Aware of the importance of the Intihuanas to the survival of the Incan people, Pizarro's men had the majority of them destroyed. But they never found the one at Machu Picchu.

Finally, an eclipse was actually to save the life of one famous explorer, Christopher Columbus, in the early 1500s. When he arrived on the island now known as Jamaica on his final voyage, the inhabitants, the Arawaks, did not welcome him with open arms. Instead they were hostile to him and his crew and refused them food. In this desperate situation, Columbus remembered an important piece of information. At sea he had been reading the newly published work *Ephemerides*, which for the first time listed the position of heavenly bodies for each day of the year. It predicted a lunar eclipse on 29 February of that year, 1504.

Columbus threatened the natives by saying he would deny them the light of the Moon if they would not help him and his crew. At first they did not believe he had such power, but, when the eclipse began as he had predicted, they begged him to restore its light. In return for food and supplies, he consented. The views of the New and Old World had come crashing together, with eclipses caught somewhere in the middle. At that time, many Europeans still clung to their traditional beliefs about the forces in the sky and it is possible that Columbus could have performed the same feat in his Genoa hometown.

But after Columbus came Copernicus and Galileo and other scientific giants. Although they had no easy time convincing people of their views, they eventually succeeded. The ancient order, along with its beliefs, was crumbling and new ideas and cultures were advancing.

---

[1] page 67: *Herodotus, I, 74*; trans. Godley (vol. I, pp. 91–93).

[2] page 68: Claudian, Gothic (or Pollentine) War, Ll 233-6, 238, 243.

Chapter Three

The Telescope

Our window to the heavens

When Italian scientist Galileo Galilei aimed his first telescope at the Moon in the early seventeenth century, our thinking about the solar system around us began to change.

The early beliefs of the ancient astronomers about the Sun, the Moon and the Earth, so carefully adhered to by government and church, began to be questioned. But change came slowly, and along the way many new thinkers, such as Galileo, were persecuted for their beliefs.

Galileo's experiments with magnification were not entirely new. Greek philosopher Aristophanes often used a glass globe filled with water to magnify his manuscripts in the fifth century BC and, in the thirteenth century, English scientist Roger Bacon claimed that the 'lesser segment of a sphere of glass or crystal' would make small objects look bigger. Branded a dangerous magician by religious leaders, he was actually jailed for 10 years.

But it was Galileo who finally used the power of glass to study the heavens around him ... and he made some of the most sensational discoveries of his time. In fact, he upset the Catholic church and fell victim to those who could not accept change.

The story of the persecuted astronomers and humankind's journey to the telescope begins around 300 years before the birth of Christ, when Greek astronomer, Aristarchus of Samos, mathematically proved that the Sun was further away from the Earth than the Moon was. He also claimed that the Earth orbited the Sun, a theory rejected out of hand by many stargazers and public leaders because it did not fit in with the philosophical, religious and mathematical beliefs of the day. At this time, scientists such, as Aristarchus, knew that the Earth was round and not flat.

Aristarchus's calculations were radically improved when Greek mathematician and

The Copernican system, showing the signs of the Zodiac, was drawn in 1660. Nicolaus Copernicus was a Polish priest who refuted the theory of an Earth-centred universe.

geographer Eratosthenes, using the varying position of the Sun as a guide, worked out the size of the Earth. It was to be the first accurate measurement of our planet with an error of less than 4,250 km. And it was Eratosthenes whose map of the ancient world was the first to contain lines of latitude and longitude. The scientific study of the Earth and our solar system had begun then, long before the telescope.

Early civilizations had always held a fascination with the workings of the sky – the patterns of the stars and the wanderings of the Moon, but mostly, they were seen as signs of pending doom or future wealth. The beliefs of astronomers and priests overlapped, with priests always having the greater say. As Chinese astronomers charted the positions of the star constellations, such as the Plough in 13 BC, the people paid homage to gods in the skies. This was a difficult time for early astronomers. They had no hard evidence that the Earth moved, for, after all, they could only see the Sun, the Moon and stars moving around it.

At around 160 BC, another Greek

German mathematician, physicist and astronomer, Johannes Kepler (1571–1630), acclaimed for his findings on planetary motion.

astronomer, Hipparchus, came to the fore with his studies of the heavens. Using a giant tube to look at the night sky, he was able to isolate stars and catalogue more than 850 of them. He also developed the mathematical science of trigonometry and advanced map-reading by improving lines of latitude and longitude.

One of the problems faced by people such as Hipparchus was the fact that the lunar year and the solar year were not the same. By the middle of the first century BC, the Roman calendar was in such a mess that Julius Caesar ordered Greek mathematician Sosigenes and his leading students to find a new system. He obliged and, although there were lots of pitfalls along the way, he thought of the bright idea of creating a leap year every four years, which sorted out the odd quarter-day difference between the solar and lunar year.

However, one astronomer's theory was believed above all others for nearly 1,500 years, even though it was wrong. Around 100 AD, the Egyptian mathematician and geographer Claudius Ptolemaeus, known as Ptolemy, produced drawings in Alexandria that showed the Earth as the centre of the universe, with the Sun, Moon and planets orbiting it while they revolved in a small circle. By using terrestrial and celestial globes, and studying and bringing together information from all the works of the great astronomers who had lived before him, Ptolemy wrote many books. One series of twelve books, the *Almagest,* catalogued astronomy; another studied astrology, a serious subject at the time, while other books dealt with such diverse subjects as mathematics, optics and music.

His writings survived only because they were translated into Arabic and were studied in the Arab lands. When the Roman Empire collapsed in the fourth century, whole libraries were destroyed and many thousands of books were burned by marauding armies and fanatic emperors.

In the early sixteenth century Ptolemy's theories were questioned. Along came a Polish priest, Nicolaus Copernicus, who totally refuted the idea of the Egyptian mathematician's Earth-centred universe.

Copernicus was born at Torun, Poland, in 1473, and went to school in Kraków and in Italy, before lecturing on astronomy in Rome. On his return to Poland, he became physician to his uncle, the bishop of Ermland, and was finally made canon (cathedral official) at Frauenburg, although he never took holy orders. He lived there until his death in 1543, mixing his astronomical work with civil duties.

Like Aristarchus, Copernicus claimed that the Sun was the centre of our solar system. His theory went against the Biblical evidence and his work was banned by the Catholic church in many countries.

Copernicus, who died in 1543, was not a great observer of the skies and spent little time gazing at them. However, he excelled in his studies of the ancient scientists, philosophers and astronomers, realizing that none of them agreed on how the universe really worked. Finally, he calculated

> ... and the moist star,
> Upon whose influence Neptune's empire stands,
> Was sick almost to doomsday with eclipse
> **Hamlet**
>
> These late eclipses in the sun and moon portend no good to us.
> **King Lear**
>
> Peace ho! the moon sleeps with Endymion,
> And would not be awak'd
> **The Merchant of Venice**
>
> O, swear not by the moon, th'inconstant moon
> That monthly changes in her circled orb,
> Lest that thy love prove likewise variable
> **Romeo and Juliet**
> **William Shakespeare**

his solar system on how long it took each planet to complete a full orbit.

His book, *On the Revolutions of the Heavenly Spheres,* was an important step forward for astronomy. Legend has it that he touched the first printed copy minutes before he died.

The task of solving the riddles of our solar system went on and, a few decades later, Danish astronomer, Tycho Brahe, came on the scene. He dismissed the Copernicus theory, saying that the planets did revolve around the Sun – but the Sun circled the Earth. In his bid to prove his theories, he made the best-ever observations of the planets, especially Mars, using the naked eye.

A particularly colourful figure of his time, Brahe had to wear a metal nose because his own had been sliced off in a duel. During his observations of the night skies, he discovered a new star – a supernova – in 1572. The star, in the constellation of Cassiopeia, shone so brightly that he could see it in daylight. Its announcement shook many people who fervently believed in the teachings of the ancient philosophers, for they ruled that stars were eternal and never changed.

The discovery brought fame for Brahe and an observatory was built for him outside Copenhagen, where he re-measured stars from Ptolemy's famous catalogue, finally publishing the world's first modern stellar atlas. But his revelations about a nova were not all that rocked society at the time. In 1577 Brahe, who was now imperial mathematician, charted a comet that moved on an orbit among the planets, disproving the Greek view that comets were only in the Earth's atmosphere.

When Brahe died in 1601, the debate about whether the Earth or the Sun was at the centre of the skies raged on and his findings now came under the scrutiny of German mathematician, Johannes Kepler, a student of the Copernicus theory. Born in Württemberg, Kepler had been Brahe's assistant since 1600 and succeeded him as Imperial Mathematician.

Using Brahe's papers and theories to aid his own calculations, Kepler worked hard to try and prove Copernicus was right. In the end, he realized that the orbits of the planets were not exactly circular, as had been believed for 1,600 years. Also, the nearer a planet was to the Sun, the faster it moved, as Mars proved. As he studied its orbit, he realized that the speed varied. At a certain point in space, Mars was travelling faster than at other points of its orbit. It seemed obvious to Kepler that the Sun was governing Mars, for, when the planet was closest to the Sun, it orbited more quickly than when it was further away.

Finally, using the shape of an ellipse (a flattened circle like a rugby ball) and not a circle, then everything fell into place mathematically. Kepler realized that the Earth and the planets all orbited the Sun in elliptical paths.

Onto the stage now stepped Galileo Galilei, the first man to use a telescope to study our solar system.

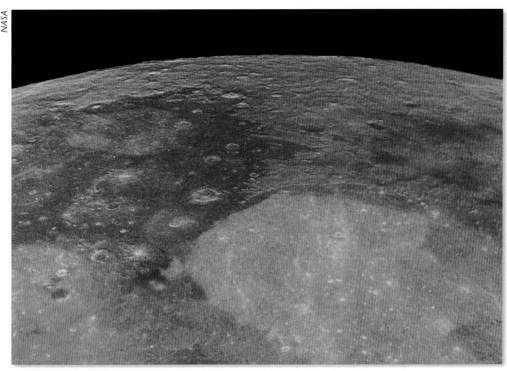

NASA

Galileo believed the dark areas of the Moon were water.
This is the Sea of Tranquillity, a vast lava plain.

Galileo, the son of an Italian musician, was born in Pisa, Italy, on 15 February 1564. He was educated at a monastery near Florence before entering the University of Pisa to study medicine. Legend has it that while in Pisa Cathedral during his first year as a student, Galileo watched a lamp swinging and realized that it always required the same amount of time to complete a cycle, no matter how large the range of the swing. It was this simple experience that helped him in his understanding of clock pendulums, mathematics and the universe.

He began to study mathematics and science with a teacher at the Tuscan ducal court, but in 1585, before he had received a degree, he was taken out of the university because of lack of funds. Returning home, he lectured at the Florence Academy and, in 1586, wrote an essay describing hydrostatic balance, the invention of which made him famous throughout Italy. Two years later, his theories on the centre of gravity in solids won him the honourable but not lucrative post of mathematics lecturer at Pisa University.

It was in Pisa that Galileo began his

National Museum of Photography, Film & Television/SSPL

Galileo Galilei and two replicas of his telescopes made in 1609, now preserved in the Science Museum in Florence. One is 126 cm long, made of wood and covered with paper. The other is 83 cm long and is made of paper, but covered in dark red leather.

These replica telescopes are recreations of a refracting telescope made by Galileo and Sir Isaac Newton's first telescope.

research into the theory of motion, disproving ancient theories that bodies of different weight fell at different speeds. Nearly 400 years later, Apollo astronauts were to take his message to the Moon, dropping a hammer and a feather from the same height in zero gravity in front of millions of TV viewers back on Earth, who watched as the feather and the hammer hit the lunar surface at the same time.

In 1592, Galileo was in quite severe financial difficulties and so applied for, and got, the chair of mathematics at Padua, where he was to stay for eighteen years.

During this time, he was convinced that Copernicus was right about the planets revolving around the Sun and often wrote to Kepler on the subject. But even though Galileo was an outspoken man, he was fearful about publicly revealing his views on the solar system in case he was condemned by the Catholic church and other astronomers of the time. That could lead to him losing his job at Padua and he had no private means of income.

Then, in 1609, Galileo was told about a Dutch optician, who had recently invented a remarkable new method of enlarging images, called a telescope. Hans Lippershey, who had a shop selling spectacles in Zeeland, had realized that by holding two lenses of different densities in a straight line and looking through them, he could make the weather vane on his local church look bigger. So he placed the lenses in a tube to keep them straight and used it to look at objects in the distance. This was called a refracting telescope because, although light usually travels in a straight line, it can be bent or 'refracted' by passing through substances of different thicknesses.

Fifty years earlier, an Englishman named Leonard Digges had constructed a simple instrument using mirrors and lenses that reflected and enlarged objects when viewed through it. This was later to be known as a reflecting telescope.

Realising the potential of such an instrument in his study of the skies, Galileo immediately started to make his own tiny lenses and mounted them in a tube. From the moment he turned his home-made telescope to the Moon, he made discoveries that were sensational in his time, and yet in reality he was using just a little spyglass that did not throw up sharp images.

His telescope's aperture (diameter of the lens), which collected light from the object being viewed and formed an image of it inside the tube, was only a few centimetres, because the technology needed to cast large pieces of glass was not available to him. Nevertheless, despite the strain on his eyesight, he discovered that there were dark spots on the Sun and four small objects orbiting around the planet Jupiter. He also realized that Venus showed changing phases just like the Moon. This meant it had to move around the Sun and not the Earth as Ptolemy had written. Copernicus was right after all. Galileo also discovered that the surface of the Moon was irregular and not smooth, as Ptolemy had thought, and he measured the shadows on the lunar surface to show how its mountains were higher than those on Earth.

Galileo became the first to be able to use the telescope to study our solar system because of the method he used for developing the curvature of lenses. He built several telescopes, grinding the lenses himself. The first only enlarged the Moon by 8 times; the last by 30 times. But it was enough to take humankind on another leap into space. Other astronomers now began to follow suit, using a lens called an object glass to form the image. A much smaller lens, the eyepiece, was fitted into the bottom of their tubes to enlarge the image, but there were still problems. The single lens in the object glass made a false pattern of rainbow colours appear around a star or planet, so the object was hard to see. Nevertheless they struggled on.

In 1611, Galileo recorded his findings, hiding them in a series of anagrams, and then took his telescope to Rome to debate the Sun and its position in the universe with the church. Leading members of the papal court were impressed and at last Galileo found the courage to make his findings public. He then wrote three letters, which were printed in Rome in 1613, maintaining Copernicus was right and Ptolemy wrong.

Galileo's claims became popular outside of the universities and created a powerful

Planet Earth

Using his telescope, Galileo discovered that Jupiter had four small objects orbiting around it.

movement of opinion among the public. However, the professors who still believed in Ptolemy united against him. They played a game of politics and their trump card was that the Copernican theory contradicted the Scriptures. Priests and preachers immediately began to denounce Galileo and other mathematicians from their pulpits. Finally, he was secretly reported to the Inquisition for heresy.

Although many members of the Church were on Galileo's side at this time, its chief theologian, Cardinal Robert Bellarmine, was not. In 1616, he could see scandal looming, which would be harmful to the Catholic Church in its struggle with the Protestant faith. His answer was to have Copernicanism deemed 'false and erroneous'. Bellarmine ordered Galileo neither to 'hold nor defend' the ideas of Copernicus, although he could

discuss them as mathematical supposition. Galileo retreated to his Florence home and quietly retired to his studies.

In 1624, Galileo returned to Rome, seeking a reversal of Ballermine's decree, and he failed to get this. But Urban VIII, a friend and protector of Galileo before he became Pope, allowed him to write about the systems of the world, both Ptolemaic and Copernican, just so long as he always concluded that human beings could not presume to know how the world was really made because God could have brought about exactly the same effects in ways unimaginable to them.

Galileo went home to his telescopes and his writing, and in 1632 he published a book about the two chief world systems, Ptolemaic and Copernican, which was received to great acclaim across Europe. It was hailed as a literary and scientific masterpiece and was studied by the Pope, who insisted on having it read to him while he was eating.

Once again however, politics and whispered intrigue came into play. The Pope was told by Galileo's enemies in Rome that the book was an outrageous defence of Copernicus, and a file was magically produced which claimed that Galileo had agreed, in his audience with Bellarmine in 1616, neither to teach nor to discuss Copernicanism in any way at all under the penalties of the Holy Office of the Inquisition. Long after Galileo's death, historians were to agree that the document had been planted.

Sunbites

- The first estimate of the Sun's distance from the Earth was made by Greek philosopher Anaxagoras (500–428 BC). He assumed that the Earth was flat and gave the Sun's distance as 6,500 km.
- The first reasonable estimate of distance was made by G.D. Cassini in 1672. He gave it as 138,370,000 km.
- Sunspots were discovered in 1610 by Galileo and Dutch astronomer J Fabricius.

The Pope had no alternative. Galileo was finally prosecuted by the church and tried for heresy. In the court at Rome, the sick and ageing astronomer, whose eyesight was failing due to his telescopic studies of the Sun, denied any memory of the 1616 document under rigorous questioning. Behind the scenes, enormous efforts were made by his friends and devotees, even church officials, to spare him – but to no avail. The man who opened a window on the solar system for the human race for centuries to come was found guilty of having 'held and taught' the teachings of Copernicus and ordered to withdraw his beliefs and findings.

On the advice of friends, an ailing, exhausted Galileo gave into the might of the church and recited a statement that he disowned, cursed and detested his past errors of believing the Earth moved around the Sun. Now he faced jail, but the Pope, through the shadowy seventeenth-century political machine, commuted the sentence to house arrest.

In December 1633, Galileo returned to his little estate near Florence, where he was to spend the last eight years of his life. Down but not broken, he worked on and his last telescopic discovery in 1637 proved that the Moon wobbled from side to side. A few months later he became blind.

Galileo Galilei died of a slow fever on 8 January 1642. Later that year, a man who was to take on and develop his ideas was born: the English scientist Isaac Newton.

Throughout his life, persecuted astronomer Galileo made one crucial scientific mistake. He strongly believed that the orbits of planets were circular, when, as we know today, that they are elliptical (Johannes Kepler had told him this). Perhaps if he had been given the backing and support of the religious leaders of the time, Galileo would have eventually come to the same conclusion.

By the time of Galileo's death and Isaac Newton's birth, scientists were desperate to study the planets in greater detail and to

National Museum of Photography, Film & Television/SSPL

Sir Isaac Newton (1642–1727). Newton's study of colours was to have far-reaching effects on the development of telescopes.

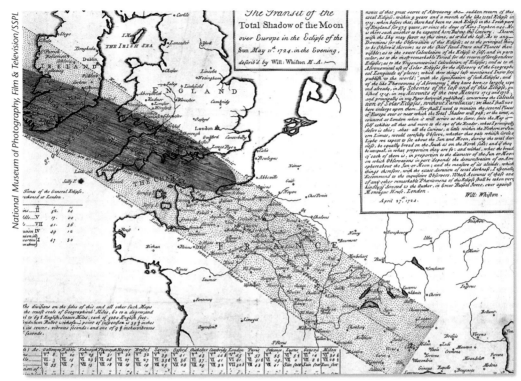

This engraving shows the passage of the shadow of the Moon over Europe on 11 May 1724. From the publication 'The Transit of the Total Shadow of the Moon over Europe in the Eclipse of the Sun' by William Whiston, MA, a clergyman, mathematician and lucasian professor at Cambridge who was expelled in 1710 for Arianism — a doctrine which refused to believe in the divinity of Jesus Christ.

focus on fainter stars. They were building longer and longer telescopes with bigger apertures. But the longer they became, the more difficult they were to handle, and so they began to strap them to poles with rope. By the end of the eighteenth century, they were as tall as the masts of ships and reclining chairs had to be designed for astronomers to look through them.

Isaac Newton was born at Woolsthorpe, Lincolnshire, and went to Grantham Grammar School, where former British Prime Minister, Margaret Thatcher, was also educated. From there, he went to Trinity College, Cambridge, becoming a Fellow.

Newton's study of colours when he was only 25, was to have far-reaching effects in the development of new telescopes and helped astronomers eliminate the problems of rainbow colours appearing around a star or planet being studied. He realized that white light could be separated by refraction

A Newtonian reflecting telescope with wooden stand, made by amateur astronomer Sir William Herschel who developed the biggest mirrors of his time.

The Men On The Moon

Alan Shepard, Jr.

- **born:** 18 November, 1923, Derry, New Hampshire
- **died:** 21 July, 1998
- **education:** Graduated from the US Naval Academy with a degree in science in 1944. Later accepted as a test pilot
- **spaceflights:** Mercury Seven and Apollo 14
- **subsequent career:** Millionaire investor

Edgar Dean Mitchell

- **born:** 17 September 1930, Hereford, Texas
- **education:** BS in industrial management, Carnegie Institute of Technology, 1952; BS in aeronautical engineering, Naval Postgraduate School, 1961; ScD in aeronautics and astronautics, Massachusetts Institute of Technology, 1964; Air Force Aerospace Research Pilot School, 1966
- **spaceflights:** Apollo 14
- **subsequent career:** Management consultant

into various colours after passing through a lens, so he decided to make a small telescope, using a mirror instead of a lens, to form an image of the object being viewed. (A mirror produced no false rainbow colours.)

Newton's idea was that the eyepiece was placed in the side of the telescope near the top. Light from the object being viewed passed down the tube and was reflected back up by a concave mirror at the bottom. The rays off the mirror hit a smaller, flat mirror under the eyepiece which reflected the object for the astronomer. But even Newton had his problems. The mirrors had to be made of shiny metal because the method of coating one side of a piece of glass with silver was unknown in those days. This resulted in the light not being reflected well – but the idea worked.

Newton published his theories about light and colours, and the motions of bodies in orbit, between 1672 and 1684 and astronomers began using his methods. Like Galileo, he was tough and outspoken, fighting the attacks on the freedoms of universities by James II. Newton even sat as a Whig Member of Parliament (the Whigs were forerunners to the Liberal party) to continue fighting for his beliefs.

Newton's love of astronomy and his thoughts on gravity never left him throughout his duties as Warden of the Royal Mint in 1695, where he carried out the coinage reform. Although his work on the telescope was to benefit generations of astronomers and scientists to come,

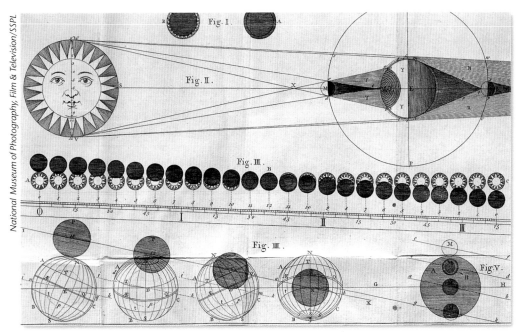

National Museum of Photography, Film & Television/SSPL

An eighteenth-century diagram illustrating the causes and results of lunar and solar eclipses.
Featured in 'Astronomy Explained upon Newton's Principles' and published in 1785.

Newton is perhaps most famous for his theory on the law of gravity. As we all surely know by now, after seeing an apple fall from a tree to the ground, he realized that such a power might extend much further, even as far as the Moon. Just like the apple, he concluded, the Moon was being held in its orbit by the pull of the Earth's gravitation.

A decade after Newton's death, a man was born who would revolutionize the telescope again – William Herschel. Herschel was a professional musician who was born in Hanover, Germany, but made his home in England. His hobbies were astronomy and mathematics, and he spent hours studying the Moon, Sun and stars from the garden of his home in Bath. He taught himself to grind and polish the biggest mirrors of his time and installed them in telescopes that he built himself.

In 1781, Herschel made a discovery that was to bring him fame: he found the planet Uranus and several of its satellites. But perhaps his most important findings were infra-red rays and hundreds of twin stars, known as binaries, which orbited each other. He also catalogued thousands of star clusters and nebulae – the vast, exploding clouds of gas from which new suns are formed – with his sister Caroline. It was Herschel, in fact, who established the basic form of the galaxy, and his discoveries

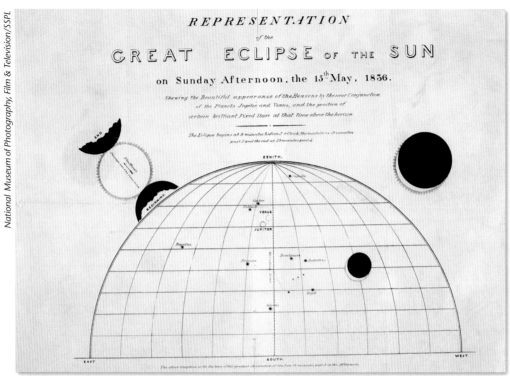

A fascinating Victorian diagram of the eclipse of 15 May, 1836, showing the positions of Jupiter and Venus, along with stars such as Castor and Pollox.

resulted in his appointment as private astronomer to George III at a salary of £200 a year.

Herschel's biggest mirror telescope was built in Slough in 1789 in the middle of the Industrial Revolution that had brought improved methods of glass-making and metal-working. It was 1.2 metres across and 12.2 metres long, and was supported by wooden scaffolding. The King was so impressed that he insisted on walking through the instrument. On another occasion Herschel and several friends not only went inside it, after an enjoyable dinner, but sang 'God Save the King' accompanied by an oboe player.

Herschel, like Galileo and the other astronomers on our journey to understand the solar system, dedicated his life to the study of the stars. He was such a passionate worker, often grinding his new mirrors for up to 16 hours non-stop, that his sister, an astronomer in her own right, who was paid £50 a year by the King to look after her brother, would spoon-feed him soup to give him strength.

After Herschel died in 1822, the next 177 years would bring even more fascinating and far-reaching discoveries.

Today, modern observatory telescopes, with their rotating domes, are built on high points far away from the glow of city lights that colour our atmosphere and spoil the view of astronomers: such places as Mauna Kea in Hawaii, 4,200 m above sea level and Cerro Pachon in Chile, about 2,715 m above sea level in the foothills of the Andes. New telescopes are currently being developed there with mirrors of 8 m across, which make objects look hundreds of thousands of times brighter than the naked human eye does.

Radio telescopes, which can pick up radio waves from the stars too dim for our astronomers to see through a telescope, have been built. They forge a path through clouds of solar dust that blocks out visible light, helping scientists to find out what is hidden at the very centre of our Milky Way.

These telescopes are usually giant bowls of metal or wire mesh that collect waves from space and bounce them into a sensitive aerial which picks up the sound. A computer then sorts out the signal from the background noise of the spinning galaxy.

Giant solar telescopes, known as coronographs, have been constructed to measure the outer layers of the Sun's atmosphere and space telescopes, such as the Hubble probe, have been sent out into the unknown.

Today, our passion to explore space – the final frontier – cannot be extinguished. As humankind now looks to the time when nuclear power may be able to take us on a voyage to Mars, the world's biggest companies plan hotels on the Moon.

Meanwhile, some 359 years after the Catholic Church condemned Galileo for claiming that the Earth orbited the Sun, he was finally pardoned by the Pope.

Moonbites

- The first telescopic map was drawn by Thomas Harriot in August 1609.
- The first British telescope observation of the Moon was made by William Lower from Wales, whose drawings have sadly been lost.
- The largest crater is Bailly with a diameter of 295 km.
- The deepest crater is Newton, with a floor 8.85 km below the crest of its wall, which rises 2.25 km above the outer surface.

Chapter Four

The Race to the Moon

What man can imagine, man can achieve

Winning the race to the Moon has been no easy mission for humankind. It has been the stuff of dreams and science fiction throughout time until this very century. But it has only been in the last 40 years that men and women have escaped the Earth's atmosphere.

Literature has entertained notions of space travel and lunar flights since the time of Christ, when the Greek satirist Lucian of Samosata wrote a parody of travellers' tales called *True Histories*, one of which tells of a voyage to the Moon by mistake. Greek mythology is full of Man's fascination with the sky. An example is the story of Daedalus and his son Icarus who donned hand-made wings of feathers and wax to fly across the ocean away from the dreaded Minotaur monster on the island of Crete. The dangers that face us in the heavens are there too, with young Icarus plummeting to his death after his wings had melted because he flew too close to the Sun, even though his father had warned him not to.

Later, scientists, such as the seventeenth-century German mathematician Johannes Kepler and fanciful fiction writers like Jules Verne, vividly pictured expeditions to the Moon. But it was not until the dawn of this century that humans created the rocket technology capable of getting the project off the ground. In order for us earthlings to leave our world behind, it was necessary to build some kind of engine powerful enough to beat the pull of our planet's gravity.

The origins of the rocket are steeped in history. It has been used in warfare since the thirteenth-century battles between the Chinese and the Mongols. Europeans used rockets in a scattering of conflicts over the following centuries, including the British attack on Fort McHenry in 1814 during the American War of 1812–14. Those flames earned eternal fame thanks to writer Francis

The first man in space. Russia's Yuri Gagarin in his Vostok 1 spacecraft. Gagarin saw what no human being had ever seen before: the Earth as a blue-brown sphere hanging in the sky against a black backdrop of outer space.

NASA

America's first manned spaceflight, Mercury-Redstone 3, designated Freedom 7, in which astronaut Alan B Shepard Jr travelled at 5,180 miles per hour. Its success prompted President John F Kennedy to pledge that America would put a man on the Moon.

Scott Key's verse 'The Star Spangled Banner' which now serves as the US national anthem.

But these rockets, just like fireworks on Guy Fawkes Day, all used gunpowder – and that would never get a man or woman to the Moon and back. It was the Russian, Konstantin Tsiolkovsky, a deaf schoolmaster who, in the early 1900s, realized that liquid propulsion would be a more effective energy source in man's mission to the stars. His ideas and findings are still applied to rocketry today, in particular the powerful combination of liquid oxygen and hydrogen is still used. Tsiolkovsky also understood that carrying such a fuel would be the heaviest part of the rocket's load – too much for a lunar flight. To solve this problem, he came up with the idea of using a multiple rocket which would shed its parts once the fuel in each one had been used up.

At the same time, American scientist Robert Goddard came to similar conclusions and, in 1926, he secretly launched the first liquid-fuel rocket with great success. But it was not long before another nation entered the arena – with dire consequences for Britain. In Germany between the wars, there was great interest in rockets and the possibility of space travel inspired much research. Under the direction of Chancellor Adolf Hitler, this was diverted to the production of the terrifying V2 rockets that bombarded London and took hundreds of lives at the end of the Second World War.

When the Allies crushed the Nazi war machine, the Americans and British took the V2 warheads and top German scientists, such as Wernher von Braun, to improve their own capabilities. The Russians were equally in the race but they only found a few discarded blueprints and many of the German technicians. It was enough, though, to start the quest for the Moon.

As if the human race's determination to stamp its footprint on the grey wasteland of the lunar surface was not dramatic enough, the crusade now became a symbolic duel between the reigning superpowers of the Cold War – the stand-off between the USA and Britain against Russia. All three countries wanted the best military and technological capabilities in the world, but all three understood the rewards of laying claim to the new frontier, the Moon and our solar system. In the passage of time, though, the Americans would eventually be the ones to stick their red, white and blue flag on the lunar surface. The British had some initial victories with their space programme centred on the Blue Streak rocket, but the cost was too high for a country still suffering from the debt of the Second World War as well as the economic uncertainty in Europe. Communist Russia, on the other hand, was determined to soldier on and had its initial victories in space.

And so the names 'astronaut' and 'cosmonaut' were born. Astronauts were American, cosmonauts were Russian. As they explored space, the world's politics and propaganda flew with them.

As it was a Russian who began work on the modern rocket, it was fitting that the Soviets had constructed the first one to leave planet Earth. In 1955, the USA started to plan Operation Vanguard, aimed at sending a man-made satellite into outer space, only to be pre-empted by the Russians 2 years later.

On 4 October 1957, the small Soviet spacesphere, Sputnik 1, which carried little more than a radio, orbited Earth in 96

NASA

America's first manned Earth-orbiting space flight, Mercury-Atlas 6, carrying astronaut John H Glenn Jr on 20 February, 1962.

minutes – and stayed in space for an incredible 3 months. Friction eventually led to its burn-out, but not before it had captured the imagination of the world, surprising and shaming the Americans. One month later, another Sputnik went up, this time carrying Laika, the first dog in space. The Russians were on a roll. Meanwhile, the USA was not faring too well. Its Vanguard rocket had exploded, leaving the Americans with red faces at the success of their red enemy. To make matters worse, Sputnik 1 had weighed a massive 83.5 kg while the Vanguard weighed in at just 1.4 kg.

Laika survived for a week in orbit and now the world knew that humans could

NASA

The first American to walk in space: astronaut Edward H White II, pilot of Gemini-Titan 4. His face is covered by a shaded visor to protect him from the unfiltered rays of the Sun.

safely travel outside the Earth's protective atmosphere. The Soviet Union was winning the space race and the propaganda war.

Then America achieved success – with the satellite launch of Explorer 1 on 31 January 1958 – all thanks to the efforts of a new US citizen, German scientist Wernher Von Braun.

But who says the number thirteen is unlucky? For the first artificial object from Earth to reach another spinning orb came courtesy of the Russians. On 13 September 1959, Luna 2 crashed onto the Moon's surface. It was a resounding success story and the world applauded. Luna 1 had missed. That same year, however, Luna 3 skimmed the lunar landscape, snapping

pictures that produced the first photographs ever seen of the dark side of the Moon.

One side of the Moon never faces Earth and, for the first time, scientists realized the two sides were totally different. The side that faced us had large, flat areas we called 'seas' because that was how Galileo saw them, the dark side had only small seas and far more small craters. Now we realized that meteor strikes and volcanic eruptions occurred at different rates on each side of the Moon, but why? Until we understand and visit the universe in millions of years' time before the death of our planet, we will never know.

The Americans were not put off by the success of the Luna probes, however. By the end of 1959, an eventful year in the space

NASA

Both sides of the Atlantic Ocean are caught by the camera of Apollo 8 and the largest, most prominent land mass is the bulge of West Africa. The portion of Africa near the equator is dark and cloudy, but the more northerly portions are clear, showing the prominent cape at Dakar and the Senegal River.

NASA

The history-making crew of Apollo 11. From left: Commander Neil A Armstrong, command module pilot Michael Collins and Edwin E Aldrin Jr, lunar module pilot. Taken in May 1969 before their Moon landing.

race, the country had moved ahead, surpassing the number of Soviet satellites in orbit. But the Americans did not celebrate for long, because in early 1961 the Russians dealt another staggering blow – they sent the first man into space.

Soviet Air Force Lieutenant Yuri Gagarin made humankind's first foray into the darkness on 12 April. The 27-year-old carpenter's son from the village of Klushinko, near Smolensk, had joined a flight training school just six years earlier, learning to fly solo and make parachute jumps, before entering the Soviet Air Force.

In 1960, he became one of his nation's first cosmonauts and was chosen from a select group of 6 pilots to complete the historic mission. Gagarin and his fellow cosmonauts, like their US counterparts, had to pass intensive physical and psychological space simulation tests. Both America and Russia required their space travellers to be in

> When first the moon appears, if then she shrouds
> Her silver crescent tipp'd with sable clouds,
> Conclude she bodes a tempest on the main,
> And brews for field impetuous floods of rain;
> Or if her face with fiery flushing glow,
> Expect the rattling winds aloft to blow
> *Georgics*
>
> Virgil

top physical condition and to have extensive flying experience, although the USA set their minimum at 1,500 sky hours, while the Russian cosmonauts averaged 250.

NASA, the National Aeronautical Space Administration, set up to take the USA into the solar system, also specified that applicants must be under 40, hold a bachelor's degree and be 1.8 m tall or less. The Russians only accepted men of 1.7 m or shorter because their spacecraft could not fit anyone bigger.

Training was tough and no wonder. When an astronaut travelled into space, the familiar background to human life was completely cut off. There was no day or night, no natural air and no gravity, and no sunshine or exercise at all.

Gagarin was cramped in his Vostok 1 spacecraft, which had a diameter of 2.26 m and was 4.27 m tall. But the cosmonaut had more pressing concerns, such as the effect of accelerating at high speeds and existing at zero gravity. He was the guinea pig for a

species – and many people made dire predictions about what would happen to him in the Great Beyond. They should not have worried.

After 90 minutes of preparation, Gagarin exploded off the launch pad at an impressive speed of 28,158 km per hour, or 8 km per second. Having been pressed back against his seat by the Earth's excessive gravitational force, he relaxed in orbit, where he hovered at a height of 180–327 km above the Earth's surface for 1 hour 48 minutes. During that time, he was able to see what no human had ever seen before and only a handful have seen since: the Earth as a blue-brown sphere hanging in the sky against the black backdrop of outer space.

Zero gravity was no problem for Gagarin, except that when he let go of his pencil after writing down his observations, it floated out of sight. The composed cosmonaut merely resumed with verbal descriptions over his radio. Although there were a few moments of panic due to a computer error and a less-than-perfect descent, he landed safely in the Russian countryside.

Gagarin quickly became an international hero and the most prominent sign of Communist achievement. He also became a source of unease for the USA, who once again lagged behind in the race to put a human being on the Moon. Sadly for Gagarin, he would not be able to claim that highest of distinctions: he died in a plane crash a few years later.

Given a jump-start by Gagarin's triumph, the Americans soon launched their own man, Alan Shepard, into space. In the air for a mere 15 minutes on a sub-orbital, up-and-down mission, his flight was captured by live television cameras which made him an instant American hero. Gagarin's mission had not been announced until it was half over. On 5 May 1961, Shepard's brief flight aboard Mercury-Redstone 3 went like clockwork, but take-off took more than 4 hours due to delays.

Although Shepard remained calm during the extended preparations, his bladder was another story. Unable to wait any longer, he relieved himself in his space suit and then radioed mission control to speed things up. When he returned to Earth, he exclaimed: 'Boy, what a ride!' There was a parade in his honour and an invitation to the White House. Following his success, President John F Kennedy pledged that the USA would put a man on the Moon and bring him back safely to Earth before the decade was over.

Sadly, Kennedy was to be assassinated and would never witness the event.

While the USA sent Gus Grissom on a carbon copy of Shepard's flight and struggled to put an unmanned craft into orbit, the Russians were launching Gherman Titov into space for a full-day mission. Only 25 years old, the pilot was the first man to sleep in space and experience space sickness. Titov travelled 702,580 km on the 1-day, 1-hour, 18-minute trip. When the USA finally sent an astronaut into orbit on 20 February 1962, 6 months after Titov's mission, the flight lasted less than 5 hours. But that mission was long enough to revive American spirits and keep the country on track in the race.

Astronaut John Glenn, who had served as back-up for both Shepard and the late Grissom, became the lucky pilot who went on the most coveted of missions: to be the first American to orbit the Earth. In his craft, Friendship 7, he witnessed the spectacular natural phenomenon of a space sunset and later saw a patch of twinkling particles he dubbed 'fireflies'. But it was not all so pleasant. While in orbit, the controls malfunctioned, forcing the Ohio-born combat pilot to do more manual piloting than he had expected. Worse yet, the ground crew received signals that the heat shield on the capsule was loose. If it broke free, Glenn could burn up on re-entry, as outside temperatures soared to 5,000 °C. Not wanting to alarm the astronaut, the controllers merely told him not to jettison his retro-pack as he had been trained to do.

NASA

A historic moment on 20 July 1969. Astronaut Edwin E Aldrin, the lunar module pilot, descends down the steps to the Moon's surface. The picture was taken by the first man to walk on the lunar surface, Neil A Armstrong.

They believed it would keep the heatshield secure. On re-entry at 24,135 km per hour, NASA lost communication with Glenn. This was normal, but in this instance, no one was sure that it would ever be re-established. Four long minutes and 20 seconds later, his voice crackled loud and clear: 'This is Friendship 7. That was a real fireball – oh boy!' He had arrived safe and sound in the Pacific Ocean and NASA's staff were in tears. Apart from cutting his hand in his eagerness to push open the hatch, he was fit and full of good humour.

After the Americans and Soviets had

National Museum of Photography, Film & Television/SSPL

HORNET + 3

Triumphant and safely home. US President Richard Nixon greets Apollo 11 astronauts Armstrong, Aldrin and Collins, in quarantine aboard the USS Hornet.

completed these basic manned-craft flights, the two nations carried out follow-up missions in preparation for a trip to the Moon, although history would record some publicity stunts.

One noteworthy mission was Russia's launch of two spacecraft in June of 1963, one carrying Valeri Bykovsky, the other Valentina Tereshkova – the first woman in space. Their assignment lasted over 3 days, pushing the Soviets' days-in-space total to 15 – or 13 more than the Americans. They also beat the USA in the gender stakes by a good 20 years because it was not until 1983 that NASA sent woman physicist, Sally Ride into the unknown. For Bykovsky and

Tereshkova, their space travel together was just the beginning of being together. The couple married 5 months later and gave birth to the world's first 'space baby'. Sadly, they later divorced.

In March 1965, 5 days before America's first launch of a manned-craft in the Gemini project – which would carry 2 astronauts and conduct trial manoeuvres in preparation for a Moon landing – Russia again stole the show. Sending their Voshkod ship up with two cosmonauts, crew member Aleksei Leonov became the first man to walk in space. Passing through a flexible airlock with a spacesuit and life-supporting backpack, he swam in the darkness at the end of a 4.6 m long cord. For 12 minutes, he performed somersaults at the rate of 10 a minute, after which he struggled for 8 more to re-enter the ship. His spacesuit, made of fifteen layers of plastic, had inflated, making movement through the narrow column difficult. He had to be careful not to tear the suit, which had cooling fluid circulating under the plastic to keep his temperature constant. But the trouble did not end there. The cosmonauts landed 965 km off course in the snowy forests of Perm, with a damaged antenna that prevented their rescue. They spent the night in their spacecraft hiding from wolves.

Regardless of technical problems, the flight broke important ground: Leonov's spacewalk proved to the world that, with the proper equipment, human beings could safely experience space outside their craft. Protected from the severe heat of the Sun and the chill of its shadow, Leonov paved the way for humankind to walk on the Moon.

That supreme feat, as we all know, was ultimately accomplished by the Americans. The Soviets, although able to capture the first pictures of the dark side of the Moon and a few other trophies, finally suffered from faulty equipment, poorly-planned missions and severe cash-cutting. They never reached the finishing line. Eventually, they turned their attention away from the Moon towards building space stations. But in the mid-1960s, the USA did not know this would be the outcome while the silence of the Cold War reigned.

At the time, the USA was concerned with

Moonbites

- The most famous plateau is Wargentin, which is filled with lava.
- The first radar echoes from the Moon were obtained in 1946 by Z Bay of Hungary.
- The first man-made probe to land on the Moon was the USSR's Luna 2, on 13 September 1959.
- The first pictures of the Moon's far side came from Luna 3 in October of the same year.
- The first manned flight around the Moon was the USA's Apollo 8 in December 1968.

merely pulling even. It did so when an American walked in space on 3 June 1965, less than 3 months after Leonov's exploit. Edward White spent 21 minutes dangling in the void, posing for pictures while using a hand-held manoeuvring device and engaging in some light-hearted banter. At one point Command Pilot James McDivitt called out: 'Move out in front where I can see you ... where are you? Move slowly and I'll take your picture ... Ed will you please roll around ... Hey Ed, smile!' Ed shouted back: 'This is fun!'

An even more important mission followed exactly 6 months later when 2 crafts, Gemini 6 and 7, were sent up to make a rendezvous. For 5 hours, the two crews chatted across a space of 15 cm, while Gemini 6 captain Wally Schirra joked with Houston controllers that they had spotted a UFO called Santa Claus and played 'Jingle Bells' in the background. Gemini 7 crew members Frank Borman and James Lovell went on to complete a 14-day mission after the encounter, and, although they emerged aching and exhausted, they proved that men could live long enough in space to make it to the lunar surface and back.

For the next 4 years, a series of developments in technology, perhaps most importantly the ability to hook up with a second craft which could convey man to and from the Moon, allowed President Kennedy's impossible dream to verge on coming true. Sadly though, in 1967 tragedies happened in both the Soviet and American camps.

Three astronauts, Virgil Grissom, Edward White and Roger Chaffee, died in a flight simulation exercise on Apollo 1, while cosmonaut Vladimir Kormarov plunged to his death when faulty machinery ruined his Soyuz 1 orbital mission.

However 1968 brought with it the breakthroughs necessary for the great adventure. Over the Christmas holiday,

NASA's James Lovell, William Anders and Frank Borman made a 6-day, 965,400-km journey to the Moon and back. In so doing, Apollo 8 was accelerated to nearly 40,225 km per hour to escape the Earth's orbit and went without ground communication for half an hour while on the far side of the Moon. The mission brought majestic views of the Earth to those gathered around television sets at home, as well as live descriptions of the Moon by the astronauts. On Christmas Day, Lovell remarked: 'The vast loneliness of the Moon up here is awe-inspiring and it makes you realise just what you have back there on Earth. The Earth from here is a grand oasis in the vastness of space.' He and the crew finished the broadcast by reading from Genesis – 'In the beginning God created the heaven and the earth ...'

Once homeward bound, Lovell quipped: 'Please be informed there is a Santa Claus.'

The dreams of astronomers, who sought to set up light telescopes and radio telescopes on the far side of the Moon to study the solar system, were drawing closer to reality. From the dark side there would be no atmosphere to interfere with viewing, no artificial Earth lights and no radio signals.

The following month after Apollo 8's success, four Soviet cosmonauts met up in space to create the world's first 'space station'. For the rest of the year, the two countries would test unmanned module landings on the lunar surface, setting the stage for history to be made.

Planet Earth

The beauty of a sunburst over the Earth, photographed from space.

Planet Earth

Shock at NASA control. The moment the Challenger exploded, killing its crew of seven.

On 16 July 1969, the sprint to the finishing line began. One million spectators and 3,000 members of the press gathered at the NASA launch site in Houston, Texas, while another 6 million people from around the world turned on their television sets to watch the momentous blast-off of the Saturn V rocket that would smash through the Earth's gravitational pull and launch Apollo 11 on its way to the Moon. Man's dream throughout the centuries was about to come true.

At 9.32 in the morning, 3 Americans –

civilian Neil Armstrong, Air Force Colonel Edwin 'Buzz' Aldrin and Air Force Lieutenant-Colonel Michael Collins – began the most daring of NASA's operations. Each man had a space flight under his belt and had earned his place on the astronaut rota for extraordinary physical and mental abilities. Yet the 3 men entrusted with this crowning achievement had no qualifications superior to those of other NASA astronauts; they were merely next in line for a trip out of this world.

The Apollo spaceship was a carefully

constructed, multi-part craft composed of the standard command and service modules and an additional lunar module, Eagle. It would be the job of the astronauts to free Eagle, containing two men, from the main ship which would stay in orbit around the Moon. The pair would then use Eagle's engine to descend to the lunar surface and, after a brief stay, power-back to rejoin the waiting ship, Columbia. It was an extremely complicated, nerve-wracking procedure carrying great risks: if they landed too quickly or an engine failed, there would be no hope of rescue.

Early on the morning of 20 July, Aldrin and Armstrong lowered the Eagle from Columbia into the darkness above the Moon. Armstrong reported, 'The Eagle has wings!' As they sped through the void, he prepared to land the craft but then realized that the terrain was filled with rocks and craters. With only 1 minute of fuel left, he looked for a better site, finding one with just 20 seconds to spare. He radioed: 'Tranquillity Base here. The Eagle has landed.' Man was on the Moon.

The astronauts were bundled up in life-protecting gear to brave elements never meant to be experienced by human beings. Temperatures ranged from 240 degrees Fahrenheit in the Sun to an incredible -279 degrees in its shadow, with no protection from solar radiation. In his spacesuit, though, Armstrong was comfortable. Six and a half hours after touchdown, he was ready to stamp the lunar surface with his footprints. He paused for a few minutes on

Sunbites

- The solar wind is the radial, continuous outflow of charged particles from the Sun. Such winds have speeds of up to 400 km per second, hurling thin matter at approximately a million tonnes a second.

- Sun particles in the solar wind are caught in the Earth's magnetic shield, called the magnetosphere, which extends into space. Two such spheres are called the Van Allen belts.

- All the colours of sunlight are contained in a rainbow: violet, indigo, blue, green, yellow, orange, red.

the descent ladder to observe his surroundings, and then stepped down, uttering his now famous words: 'That's one small step for a man, one giant leap for mankind.' The world applauded in agreement. NASA was ecstatic. Television

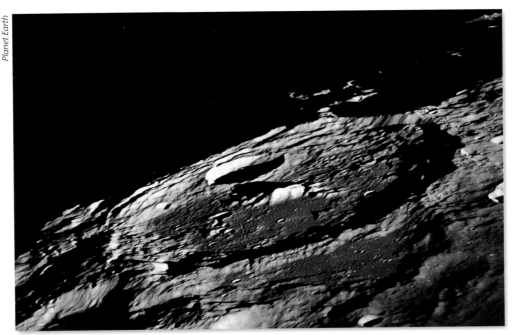

Planet Earth

The barren lunar surface photographed from Apollo 13.

viewers on Earth were mesmerized by Armstrong's flickering black and white image on the lunar surface, recorded by a camera he carried with him.

Aldrin joined him describing the scene as 'magnificent desolation'. The pair then read the inscription on a plaque they were to install on the lunar surface: 'Here men from the planet Earth first set foot upon the Moon. July 1969 AD – We came in peace for all mankind.' They also left an American flag, a memorial to the deceased astronauts and cosmonauts of the Space Race, and chatted with President Richard Nixon over their radio link.

'This certainly has to be the most historic telephone call ever made,' the President declared. After collecting samples for more than 2 hours, conducting experiments and cavorting on the grey, rocky surface, Aldrin made a few kangaroo hops in the low gravity before they climbed the ladder back into the Eagle, where they spent the night before taking off and being reunited with Columbia and Collins.

The world was told that their footprints would remain intact on the lunar surface for billions of years to come as there was no air and therefore wind to blow them away – unless another spacecraft blasted the dust.

A few weeks later, rocks from the Moon were being studied by geologists. They were ground down into thin slices and put under powerful microscopes. These samples contained more of some elements such as titanium and less of others such as gold. The ages of the rocks ranged from 3.1 billion to 4.5 billion years.

Over a decade in the making, with 12 billion pounds spent and 3 lives lost, the United States of America had won the space race. Humanity had won the Moon.

The saga of manned space flight of course, does not end with this triumph. Although the first voyage to the Moon gave scientists valuable information about its components and properties, there was much more to learn. The USA went on to make 5 more lunar landings, even sending a geologist along. The Soviets continued to shoot for the Moon, although they eventually had to be satisfied with samples which were collected by robots, officially ending their quest for a manned lunar landing in 1974.

Meanwhile, NASA built land rovers to help future astronauts explore more of the lunar plains on each visit.

In July 1971, astronaut Dave Scott took the message of Galileo to the Moon in Apollo 15. As he stood on the lunar surface, he told a worldwide television audience that the persecuted Italian astronomer was right after all. 'Gravity pulls all bodies equally regardless of their weight.' Scott then dropped a hammer and a feather from the same height and watched as, with no atmosphere to alter their fall, they hit the surface at the same time. It was a spectacular mission, providing continuous colour television coverage.

Planet Earth

The Eagle has landed: Armstrong's first footprint on the Moon's surface.

The USA's final Moon trek was made by Apollo 17 in December 1972. It lasted 22 hours and ended with another plaque being placed on the lunar surface. It read: 'Here Man completed his first explorations of the Moon in December 1972 AD. May the spirit of peace in which we came be reflected in the lives of all Mankind.'

Both the USA and the USSR now turned their attention to space stations, and in 1975 bilingual astronauts from both countries shook hands in space.

In 1981, Columbia, America's first Space Shuttle was born, heralding a new era in space exploration. The Shuttle could take off like a rocket and land like a glider. It could be used time and time again, ending the expensive practice of using rockets only once – The Americans were streaking ahead once more.

Planet Earth

American astronaut James Irwin on the lunar surface, Apollo 15, 1971. Fellow astronaut Dave Scott took Galileo's theory to the Moon when he dropped a hammer and feather from the same height and watched as, with no atmosphere, they hit the surface at the same time.

In 1986, the Russians launched the Mir space station, and the following year it became the first to be continuously inhabited. Later it served as a site of joint operations between the two countries.

But 1986 was a bad year for America. Tragedy struck when its Challenger spacecraft exploded and 7 astronauts died, including schoolteacher Christa McAuliffe.

The mid-1980s also saw the first black astronaut, mission specialist Guion Bluford, head for the stars, along with physicians and congressmen. Medical tests that were to assist in the new world of heart transplants were undertaken along with several other studies and experiments that were to help man survive and enjoy a better life on Earth.

More recently, NASA has turned its attention towards the planets and has sent probes out into the solar system.

Though we have been to the Moon, we still want to know more about the universe around us. Many people believe that we will soon colonize the lunar orb in our skies and visit other worlds. These ideas are not as far-fetched as they were in the time of Galileo. Our current fantasies now appear to be the reality of the new millennium. If we can do what we have done in 50 years, what will we achieve in the next 1,000?

America's Shuttle spacecraft are already involved with Freedom, by far the biggest-

Planet Earth

What Galileo couldn't see through his telescope: The vast emptiness of the lunar landscape.
(Taken from Apollo 17)

ever permanently manned space station under construction.

We must always remember, however, that mankind did not arrive on the Moon through the efforts of the last five decades alone. It has been through the efforts of astronomers, scientists and politicians throughout all of human history.

Telescopes and breakthroughs in rocketry, computers and engineering were all necessary to make Apollo 11 the success that it was. But so was the continual desire of men and women to seek, to learn, to conquer and to understand the universe around them.

Our intelligence which eventually created a sophisticated flying machine, our sense of adventure which allowed Yuri Gagarin to climb into a tiny silver shell and be blasted hundreds of miles per minute into the darkness are all part of the story.

But, above all, the cleverness of mankind is our ability to learn from past mistakes. More than any one astronaut or institution or country, the entire human species ultimately deserves credit for getting us to the Moon.

Even the legendary Greek Daedalus and his son Icarus fleeing from Crete have played their part.

The Men On The Moon

David Randolph Scott

- **born:** 6 June 1932, Randolph Air Force Base, Texas
- **education:** BS in military science, United States Military Academy, 1954; MS in aeronautics and astronautics, Massachusetts Institute of Technology, 1962; Air Force Aerospace Research Pilot School, 1963
- **spaceflights:** Gemini 8, Apollo 9, Apollo 15
- **subsequent career:** President, Scott Science and Technology, space transportation and exploration development consultant

James B. Irwin

- **born:** 17 March 1930
- **died:** 8 August 1991
- **education:** BS in naval science, United States Naval Academy, 1951; MS in aeronautical engineering; MS in instrumentation engineering, University of Michigan, 1957
- **spaceflights:** Apollo 15

A home in space

While the age of the space race got its name from the competition between the two superpowers to be the first to set foot on the Moon, the current era of space exploration is equally characterized between efforts at cooperation between Russia and the USA. Today, NASA's flagship programme is the construction of the International Space Station, spearheaded by the 2 countries in conjunction with 16 others, including the 11 nations of the European Space Agency.

The space station is the most ambitious international scientific venture ever undertaken. This massive structure will be larger than Washington DC when it is completed in 2004. It will feature an acre of solar panels alone in order to power the craft's 6 laboratories. While the USA will provide a large bulk of the facilities, other countries are involved in constructing components and contributing equipment. The European Space Agency, for instance, is creating a pressurized laboratory and space vehicles. Ultimately, the station will function as a research centre to advance our understanding of the solar system and pave the way for future space travel.

Reams of experiments and observations form a fundamental part of the station's mission, ranging from studies of human behaviour to investigations of space, to probing areas of financial benefit. Specific goals include: identifying the effects of life at zero gravity on living matter; collecting more information about the composition of elements in space and their effect on

NASA

Russia's Mir Space Station against the Earth's horizon, photographed from the US Space Shuttle Discovery. Mir and Discovery heralded a new era in space exploration.

spacecraft; increasing the understanding of long-term changes in the environment and the effects of air pollution.

The joint USA–Russia Shuttle–Mir Programme has already laid the groundwork for the international space station. Although Mir, which began in 1995, has been plagued by equipment failures and facility breakdowns, it has enabled astronauts to experience more than 2 years of continuous stays in space. In comparison, it took the US shuttles 12 years and 60 flights to log just 12 months. In 1998, the first 2 space station components were launched: first the US-owned, Russian-built Zarya on 20 November and then the US-built Unity module on 4 December. Over the next 6 years, the remaining components will be added through additional missions and assembly will take place in space.

The first crew to live in the station will not be installed until the year 2000, following a Russian Soyuz spacecraft launch. American commander Bill Shepherd will be joined by the Russian Soyuz Commander Yuri Gidzenko and Flight Engineer Sergei Krikalev. They and other crew members will then participate in a battery of space walks and assembly missions.

But it is not the only project NASA has on the burner. The Hubble Space Telescope, whose photographs of faraway galaxies and previously unseen parts of the universe have attracted attention around the world, continues to gather its celestial data. In the meantime, a high-tech spacecraft continues its voyage to the moon Europa, half a billion miles from Earth, to investigate whether it contains oceans underneath its icy surface. Closer to home, scientists recently found evidence that the poles of the Moon might themselves contain billions of tons of ice.

NASA's newest missions, set to begin in the fiscal year 2000, will pay special attention to Earth's neighbouring planet, Mars.

We have come a long way since the ancient astronomers, but there is an even longer way to go.

NASA

On 28 January 1986, the crew of the Space Shuttle Challenger lost their lives following an explosion during the launch phase. Left to right, front row: astronauts Michael J Smith, Francis R (Dick) Scobee and Ronald E McNair; Ellison S Onizuka, Sharon Christa McAuliffe, Gregory Jarvis and Judith A Resnik. McAuliffe was a payload specialist representing the Teacher in Space Project.

Tragic take-off

Millions of pairs of eyes around the world watched with anticipation, admiration and ultimately horror as the Challenger suddenly transformed itself from a sturdy rocket into a deathly fireball 73 seconds after take-off. Many of those eyes belonged to children. This was the space launch for the young generation that was meant to encourage them to study the sciences and get them excited about a space programme that was by then old news. It was the mission for which one lucky teacher had been chosen from 11,000 applicants to be the first 'average person' in space and, as such, to become every child's teacher. This was the event that schools around the country marked by gathering their students in auditoriums before televisions to share in the excitement.

Yet the day had dawned so brightly – too brightly in fact. After a host of weather delays due to dust storms and drizzle, the sky on 28 January 1986 was a crystal-clear blue.

Although the freezing cold caused icicles to form on the Shuttle, NASA decided to go ahead with the already behind-schedule launch. A thousand spectators crowded the stands for this dramatic moment, as McAuliffe and her 6 accompanying astronauts waved to the crowd and bravely strode towards the launch pad. The usual reporters and photographers attended the launch. Also present were McAuliffe's parents, husband, 2 young children, and 18 of her third-grade students from Concord, New Hampshire. They knew, as they watched the young woman enter the spacecraft, that history was being made and her life would never be the same: They did not know the awful reason why.

The launch itself went off without a hitch and, for the first minute of the expedition, everything went like clockwork. The crowd cheered enthusiastically as the rocket blasted off into the blue beyond with all the power and elegance of humankind's finest machinery. Inside the ship, the crew went through rounds of systems checks, all of which were reported on track. The conversation between the astronauts and the ground crew even up to the last seconds of the doomed mission remained eerily normal. It was only on the television screen that anyone could catch the small orange flame escaping from the flank of the spacecraft, and even that only flickered for an instant before the Shuttle exploded in two long streams of billowing exhaust and bursts of fire. Yet, to the inexperienced viewers on the ground, it was still not clear that anything had gone wrong. At their angle, the explosion appeared similar to the planned separation of the rocket's boosters. When the terrible truth became all too clear, family members and students in the stands began to sob, and the sound of mourning echoed around the world. A mission meant to captivate the country's collective imagination and bolster public regard for the space programme instead unified society in a moment of shared tragedy.

In the months following the disaster, its causes slowly unfolded, revealing a chain of bad decisions and overlooked safety risks on the part of NASA officials. The Shuttle, on its 25th mission, had already sparked off concerns that it was no longer safe for flight, while the frigid weather raised another red flag. The engineers who helped design the rocket claimed that they had argued against the launch, particularly because of the low temperature; while NASA officials claimed their objections had not been that vehement. To many observers, it was a disaster waiting to happen, the question being not whether, but when. For the Challenger's most publicized mission, bad judgement, political pressure and public expectations helped override pleas that it be grounded. The results were disastrous, but in the end, also instructive. Although the mission had intended to teach happier lessons about our domination of the final frontier, it still left the world with important lessons: life is fragile, travel in space is dangerous, and those who would accomplish one must risk the other. That is the basis of human progress. ✺

Chapter Five

The Eclipse Chasers

'You can't describe it, you have to see it'

One of the most fascinating and awe-inspiring sights known to man and woman – a total solar eclipse – is the passion of a group of people who spend a great deal of their lives chasing and capturing this natural phenomenon on film. They are the eclipse chasers, a band of dedicated photographers, scientists and astronomers who travel the globe to experience the mighty solar spectacle. Here are the stories of three men who follow the Sun.

Michael Maunder has survived snowstorms, altitude sickness and rough days at sea in his treks from one eclipse to another. Yet no matter how severe and dangerous the conditions, how long the journey or how short the instant of totality, he has never regretted going. 'Every total solar eclipse is different and they are all absolutely out of this world,' he says with a smile. 'They are so rare that England hasn't seen totality since 1927, although other

nations have. And, after 1999, the English will not see another one until the year 2090, when most of us will be gone.'

Just 2 years ago, at the age of 59, Michael and his wife Wendy travelled to the icy, winter plains of Mongolia to take pictures of a total solar eclipse over a frozen lake. But they never reached the site because their bus was bogged down in a blinding snowstorm. Luckily for the passengers, the weather was warmer than expected, hovering around 5°C below zero but there were fears that it could drop to –25 °C – life-threatening level. As the snowstorm cleared, clouds covered most of the sky, and, with only ten minutes to go before totality, their destination, the frozen lake, was not even in sight. Everyone was in despair after travelling thousands of miles to reach the base they had dreamed about for twelve months. Intrepid Michael would not be deterred, however. He jumped out of the

Michael Maunder sets up his camera for totality. (Montana 1979)

bus and set up his cameras with the special, filtered lenses on the side of the icy road. He only had time for a few wide-angle shots of the eclipse as the Moon blocked out the Sun, but the results were fabulous. 'When I look at those photographs now, there is a serene stillness, an isolated fragility to them. The darkened Sun sitting in the middle of those snow-covered hills gives me a feeling of peace and tranquillity and an understanding of how beautiful nature is,' he says. 'It was a breathtaking experience, but it wasn't a place I would ever go on holiday. No way. And yet I now have happy memories and hope to return someday.'

Michael is the first man to tell people about the adventure and beauty of experiencing and photographing an eclipse – and the first man to warn them of the dangers. 'The Sun appears beautiful when it is low over the horizon and shining through a layer of haze or mist,' he says. 'Somehow it looks safe when it is not glowing bright. But it is always dangerous even during an eclipse – and especially so for those people like me whose mission is to take pictures of it. Everyone should remember never to stare straight at it, never to view the Sun through

an unfiltered telescope or binoculars and [always to] seek help and advice about lenses when using a camera. There is so much to learn – and not just about safety. Camera shake, for example, can ruin a photographer's work at a vital time and they may only have a minute to capture the wonderful spectacle of totality. There is no way a photographer can stand back and shout, "Can we try that again please!" '

Many of Michael's eclipse adventures have taken him to wonderful holiday destinations, however. In July 1991, he went to the paradise islands of Hawaii, where he travelled miles from the hotel on a coach blessed by hula-hula girls, to a site up in the hills offering the best view of the total solar eclipse. As he set his cameras up, the grey clouds swept over him and so did despair. The pictures of the Moon and the Sun he had been planning were impossible. Ironically, if he had stayed at the hotel, he could have photographed the eclipse properly in a nearly clear sky overhanging the Pacific Ocean, although clouds were everywhere that day. That's the risk people like Michael take: planning for months or even years for the great moment, only to see their dreams collapse in a few seconds, thanks to a few puffs of grey cloud.

But not all of his expeditions have been so troubled. On one occasion, in 1976, he and his travelling companions went to the only spot where the total eclipse was clearly visible in the world – Zanzibar. A local drove the rickety old bus across the sandy plains and bumpy roads, while Michael and his friends wondered if he knew what he was doing or where he was going as drivers were required to report to each local police station on route. All they knew was that Bill Thomas , an amateur weather forecaster and Liverpool bus driver, had calculated 18 months in advance the one beach front that would offer a clear view of the event. But his prediction was so accurate that they all got a series of clear photographic shots, while 4 miles up the road, the President of the country, who arrived with marching bands for the occasion, was clouded out.

Catching an eclipse is not only a matter of choosing the right place, but also the right type of transportation and equipment. Although an 'eclipse virgin' on his 1973 trip to Mauritania, Michael took along some sophisticated equipment which overcame the difficulties of taking pictures on a swaying boat.

To get there, he and 350 other enthusiasts travelled 4 days by sea from Liverpool. Out on the water to view the eclipse, Michael mounted his camera on a home-made structure resembling a rifle butt to minimize the effect of the rocking from the waves. An experienced shooter, he knew how to hold the instrument and the 'rifle butt' steady against the motion and, as he gripped, he prayed. From those 6 minutes of

India, October 1995: Michael Maunder uses a wide-angled, fish-eye shot to capture the sloping horizon during totality. (Michael Maunder)

Roses have thorns, and silver fountains mud;
Clouds and eclipses stain both moon and sun,
And loathsome canker lives in sweetest bud
Sonnet 35

Not mine own fears, nor the prophetic soul
Of the wide world dreaming on things to come,
Can yet the lease of my true love control,
Suppos'd as forfeit to a confin'd doom.
The mortal moon hath her eclipse endured,
And the sad augers mock their own presage;
Incertainties now crown themselves assured,
And peace proclaims olives of endless age
Sonnet 107

William Shakespeare

totality onwards he was committed. Amazingly, the pictures he did not think would be perfect were magnificent. 'No matter where you are, even at sea or in a dangerous situation, a total eclipse is an awesome, overpowering, impressive thing – and you must hold your nerves and excitement when you try and take pictures of it. It is an art form and the reward is having those photographs forever. No wonder I am hooked. A total eclipse is addictive – I am always thinking of the next one,' he says.

That explains how two decades later, in November 1994, he travelled 4,267 m above sea level to a mountaintop in the Chilean Andes, battling against the debilitating effects of the thin air. It was to be the most impressive eclipse that he had ever photographed and experienced. 'It was almost a religious experience,' he says. 'I stood staring through my camera lens with my heart beating as the Sun's corona during the eclipse burst out in fiery streaks, with two flowing upwards and one down in a vision, or an eerie face in the heavens with horns,

Michael Maunder

Timed sequence of the eclipse seen from Guadeloupe in February 1998.

just like the Devil. Set against a background of mountain peaks and smoking volcanoes, it was chilling. I suddenly understood how the ancients thought the eclipse was a force of evil.'

Michael certainly had to fight off devilish deterrents to get there – one third of the party stayed behind at the eclipse party's camp because they were suffering badly from altitude sickness; his wife Wendy decided not to make the trip to Chile at all even though she was disappointed about not going; and the ascent to the site took the whole night, with Michael's bus leaving the base at midnight to arrive at dawn. Breakfast of wheat flakes and yoghurt was at a military camp – the only place in the region capable of feeding 300 people in a sitting – and then the army escorted the eclipse chasers to the summit because the government feared raids on them by drug traffickers and machete gangs along their path, which sat on Chile's borders of Bolivia and Peru.

At the top, Michael felt better than most and set his cameras up perfectly. But the eclipse was so spectacular that, after capturing it through his lenses, he became so excited that he started running around. Moments later he collapsed feeling sick and breathless. 'You can't get too excited up there,' he says. 'I learned the hard way.'

But his excited reaction, he discovered, was a normal one. He later attended a solar talk, during which the lecturer described an expedition on which he was accompanied by 6 professional astronomers who had never seen an eclipse. He had taped their responses and played them back for the audience. 'It sounded as if there were scores of people, because of all the clamour and commotion. It was a real din,' says Michael. 'But I understood how they felt.'

Michael will view the 1999 eclipse from his back yard – literally. He lives in Alderney, the only Channel Island in the path of totality. Alderney, which has a population of 2,500, is expecting 10,000 visitors. The Maunders will be joined by, among others, a team of professional astronomers on a hill a few yards from their home. The summit sits at the top of Fort Albert, a Victorian site which also includes their house. The view from the top offers a 360-degree horizon, a key factor in observing effects like the waxing and waning of the Moon's shadow.

Eclipse chaser Michael is not a professional astronomer. He did not study the subject at all in school. His interest in the phenomenon was sparked off by a partial eclipse he saw as a 10-year-old boy. During his secondary school days, he saw a string of others and he still remembers the day when astronomer Patrick Moore came to set up a telescope in the playground. Today they are friends and have even written a book on solar eclipses together. Now he wryly sighs that a Boots pharmacy sits on that very site. He did, however, join an astronomical society and pursued his interest at an amateur level while studying other sciences and taking a degree in chemistry. Michael is a chemist – of a special sort. He makes chemicals which quickly develop film, some of which have helped

Michael Maunder

Zanzibar 1976: Michael Maunder prepares for the eclipse.

him improve his own eclipse photographs.

Michael happily notes that this mingling of astronomy and photography, even on an amateur level, can have significant results. Pointing to the work amateurs have done to discover meteors, supernovae and other astral phenomena, he believes that amateurs and professionals are increasingly working hand in hand because astronomers do not have time to look for everything. He proudly explains that the observations of eclipses made by himself and his friends, who all have cameras focused on the same part of the sky from different places, have offered scientists the first accurate UK measurement

of the speed and orbit of a meteor. 'A lot of science can come out of one little picture,' he says.

The total eclipse of 1999 coincides with the annual meteor shower of the Perseids, which peaks on 11 August. Michael and his eclipse-chasing colleagues hope that they may be able to capture the total eclipse and a meteor in the same frame.

But much as the idea of such a picture – and the rest of his collection of photographs – excites him, Michael is the first one to admit that experiencing an eclipse through photography is not enough. 'If you haven't seen one before, you are totally unprepared

Michael Maunder

*The breathtaking beauty of third contact, around 4,200 m above sea level
in Chile in 1994. Venus is seen top right.*

The Men On The Moon

John Watts Young

- **born:** 24 September 1930, San Francisco, California
- **education:** BS in aeronautical engineering, Georgia Institute of Technology, 1952; Naval Test Pilot School, 1959
- **spaceflights:** Gemini 3, Gemini 10, Apollo 10, Apollo 16, Space Shuttle STS-1, STS-9/Spacelab 1
- **subsequent career:** Special Assistant to the Director for engineering, operations and safety, Johnson Space centre

Charles Moss Duke, Jr.

- **born:** 3 October 1935, Charlotte, North Carolina
- **education:** BS in naval science, United States Naval Academy, 1957; MS in aeronautics, Massachusetts Institute of Technology, 1964; graduated from the Air Force Aerospace Research Pilot School, 1965
- **spaceflights:** Apollo 1666
- **subsequent career:** Christian ministry

for totality,' he says. 'Even if you've seen it on a film or on the television, you have to experience it. You can never adequately reproduce how the light drops; you can't even describe it. No picture will do it justice.' For him, the best part is the diamond-ring effect, when the first drop of sunshine emerges from behind the Moon as the Sun comes out of the eclipse, creating a glowing circle of light at the circumference. Although it lasts a mere fraction of a second, 'that single phenomenon is so exquisitely beautiful that it's worth going to an eclipse just for that event. It is so exquisite you can't describe it. You have to see it,' he says. 'The moment is so fulfilling.'

Another eclipse chaser, who has helped us to produce this book, is Francisco Diego. 'An eclipse is an elusive thing – you have to follow it wherever it goes, you have to be chasing it, you have to travel long distances to catch up with it – and when you get to it what you have come to see is over in just a few moments,' he says.

Thirty-seven-year-old Francisco is a senior research fellow for the department of physics and astronomy at University College, London. Nicknamed Cisco by colleagues, he has a PhD in astronomical instrumentation and is well known for his studies of total eclipses, having led 10 world expeditions to experience the solar phenomena.

His newly developed instrumentation has enabled him to obtain valuable photographs of the Sun's corona and chromosphere, highlighting the elusive diamond-ring effect, studied by scientists for centuries. Francisco's pictures have now been put together for a breathtaking calendar that takes the viewer through each stage of a total solar eclipse, including awesome shots of the landscape.

Francisco did not have to travel far to see his first total solar eclipse. It took place in his native Mexico when he was 20 years old. 'The sky was completely clear, it was perfect for the event and I couldn't believe my luck,' he recalls. 'Imagine it, I was able to see my first eclipse with Baily's Beads, the incredible diamond ring and shadow bands in a fantastic way. It was everything that I had hoped and I was spellbound.'

Since those precious few minutes, Francisco has been to Quebec, India, Bolivia and many other places to study and experience the phenomenon. In 1977, he even spent 10 days aboard a US Navy frigate in order to view an eclipse in the middle of the Pacific Ocean. 'When the Moon began to eat into the Sun and I could see it clearly, I knew that this would be a fantastic and almost chilling memory,' he recalls. 'When totality came I was filled with wonder along with all the American sailors. It seemed as if they were in silent tribute to the universe. It was very impressive.'

Although Francisco has chased so many eclipses and recorded them throughout the world, he remembers each one with ease due to their distinguishing characteristics. In 1980, he saw a 'spectacular' eclipse from the middle of a national park in Kenya. 'The corona was very bright and extended in all directions because the Sun was at the

Michael Maunder

The miracle of the Sun. A volcanic ash sunset in Namibia captured through the lens of Michael Maunder.

maximum of activity,' he says. 'There were no people around me, it was all silent and I felt so privileged to experience it this way. It made me feel very small but helped me understand even more how little we know and how much more we must learn. It made me want to take this message to people going about their everyday lives, so that they might understand the universe around them. That day, even some cloud, which sometimes spoils the occasion, made the event and my pictures even more dramatic.

'Apart from the beauty of such a spectacle, you get a very, very special feeling when you are in the shadow of the Moon and you can see the limits of the shadow all around you. It is as if the Moon is pointing a finger at you.'

For Francisco, there is more involved in observing eclipses than the feelings of awe they invoke. 'It's not only the beauty and the wonder this inspires, but the science of it,' he says. 'Part of the work we do at the university is with solar physics and the way the Sun behaves. Solar eclipses are very important ways for us to find out about our local star, which through its gravitational pull holds us and the other planets in orbit around it. Observations from the ground of an eclipse are unique – even a satellite can't do the same observations.'

During an eclipse, the Sun's thinner outer layer, the corona, is finally visible, and this allows Francisco and his fellow scientists to conduct experiments giving greater insight into the star's composition and structure. He can see the way in which the

Nick Quinn

Nick Quinn. 'Anyone who hasn't seen an eclipse, can't look at a photograph and know what it is really like,' he says.

corona moves and behaves, and how the matter in the corona is flowing. 'That's so important because our planet is immersed in the solar corona so we have to understand it,' he says. 'The air, the weather – the things that give our planet its unique atmosphere and our life – come from its properties.'

Francisco hopes that the magic of the Cornwall solar eclipse will help spark enthusiasm for astronomy in particular and the sciences in general. His mission is to take the Sun, the Moon and the Earth to the people in the form of planetarium shows and lectures across Britain, using the eclipse of 1999 to encourage more youngsters to

Nick Quinn

The diamond-ring effect, third contact 3 November 1994, seen from Chile and caused by the last rays of sunlight passing between mountains on the lunar limb. Nick Quinn also captures prominences, seen as red dots in the pearly-white inner corona of the Sun.

Nick Quinn

The Sun's corona can only be seen during a total eclipse. This picture was taken with a telescopic camera in Chile in 1994.

study the workings of the universe.

Francisco also hopes the experience will bring people back to nature and inspire a greater interest in acquiring knowledge for knowledge's sake. 'Society is moving away from nature,' he says sadly. 'More and more we are living enclosed lives in our cities – rushing around doing our own business and thinking about money, health and politics. People do not have time to relax and really learn things for the sake of knowledge. The eclipse can change that. Important events like this can produce the pleasure of discovery and the satisfaction of a fundamental human characteristic – curiosity. All people need to do is learn how the universe works. The purpose is that knowledge itself, and not the application of that knowledge.'

Eclipse chaser Francisco became interested in astronomy when he was a child and still has his first telescope – a fifteenth birthday present. From the moment he unwrapped it, he used his telescope to make observations and sketches of the night sky. Twelve months later, he joined his local astronomical society and, because of the satisfaction and help given to him, he is still a member today. But his first degree was in mechanical engineering, while astronomy – and constructing telescopes, just as with Galileo and Herschel – remained a hobby.

'Little by little I joined astronomy professionally,' he says. 'I just couldn't help it after experiencing an eclipse. Finally, I got a job at a university in Mexico, building

Nick Quinn

Totality in Chile in 1994.

telescopes, and then came to London, where I got my PhD.'

But even with his advanced learning and professional interests in astronomy, Francisco, like so many others, cannot help being overcome by the most spectacular event in nature – the total eclipse he has witnessed so often. 'You feel a bit of terror and fear shoots right through you, even though you know what is happening,' he says. 'The adrenalin pumps through your body as the giant shadow of the Moon sweeps over the Earth. In our own background, our past, we still have an inherited fear of this event. It has been passed down from century to century. The sudden disappearance of the most

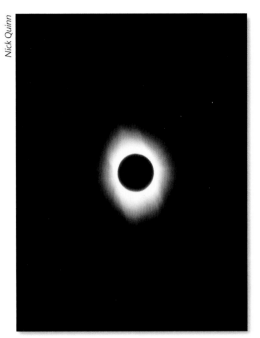

Nick Quinn

Nick Quinn

Totality in Venezuela. 26 February 1998 ...

... and again, but using a longer exposure.

important thing in our lives, the thing that gives us life, the Sun, is a very chilling experience. No wonder ancient societies feared eclipses and believed them to be the wrath of the gods. These forces which were incomprehensible, still are incomprehensible when you witness them. We know the mechanisms that drive them, but we still don't understand them.

'When it comes to eclipses, you feel nervous. I have stood in the shadow of the Moon many times, and the feeling that something special is happening never goes away. And yet every time there is something different about it.'

Amateur astronomer Nick Quinn's passion for total solar eclipses has led him halfway around the globe. His first encounter was on the Indonesian island of Java, where he saw a sight which, 16 years later, he still cannot find adequate words for. 'It was simply awe-inspiring,' he says. 'I don't know how to describe it. I was hooked. From then on eclipses became addictive because they are so short. You only get a few minutes at a time, so you think to yourself – I must see a few more! I must, I must!'

Nick went on trips to locations as diverse as Mexico, India and Finland. 'Each one was different,' he says, hesitantly choosing Chile as his favourite eclipse destination. There he and his band of solar chasers had to travel to the top of a mountain 4,267 m above sea level to experience and photograph the awesome spectacle. Even though he felt sick and listless from the effects of the altitude after the 5-hour drive up the side of the mountain, the view – overlooking the Chilean plains and a volcano – was stunning. 'I will never forget it,' he says.

Although the quiet and isolation made this experience stand out, Nick believes that sometimes it is the reactions of the thousands of onlookers around him that makes the spectacle even more memorable. A total eclipse over the beaches of Venezuela in 1998 was an historic moment for the country. Thousands of people in the path of totality sang their hearts out and shouted in almost hysterical celebration as shadow bands swept around them. The occasion was electric. People were in tears and on their knees when the Sun came back into view as if the warmth from the skies gave life again.

Other spectators across the globe are less enthusiastic, however. Even today, some people still believe eclipses are a sign that the end of the world is nigh. When Nick was in Java, the government told pregnant women to stay in their homes during the eclipse so that they could still bear children in the future. And, at the 1994 annular eclipse in Morocco, when Nick and his fellow solar chasers started talking about the impending event to locals, they fled in fear.

Placing a protective filter in front of a telescopic camera lens enables Nick to take photographs of the natural phenomenon that always fascinated him. This technique came naturally because his early hobby had

Moonbites

- The first man on the Moon was American Neil Armstrong, in the USA's Apollo 11. He stepped onto the lunar surface on 21 July 1969, followed by Edwin – known as 'Buzz' – Aldrin. Two years later, on 30 July 1971, US astronauts Randolph Scott and James Irwin drove their lunar roving vehicle (LVR) in the Apennines area.

- The first professional geologist on the Moon was Harrison (Jack) Schmitt in the USA's Apollo 17. On 11 December 1972, he found the famous orange soil, nicknamed 'shorty'. The colour was due to small, glassy particles, which were ancient solar matter of around 3.8 billion years old.

Sunbites

- Prominences were first described by the Swedish observer Vassenius at the total eclipse of 1733, although he believed they belonged to the Moon and not the Sun.

- The first radar contact with the Sun was made in 1959 at the Stanford Research Institute, USA.

- The first attempt to show a total eclipse on television from several stations along the path of totality was made by the BBC on 15 February 1951. It was successful and totality was shown successively from stations in France, Italy and Yugoslavia.

always been photography. Combining astronomy and photography made sense to him from an early age. Although he only studied astronomy formally in secondary school, he has had a passion for space since he was a child, growing up in the midst of the Apollo Moon landings. Today, he still vividly recalls 'the flickering black and white image' of American astronaut Neil Armstrong stepping onto the lunar surface and making history as the first man on the Moon. Such an event, he says, was 'quite difficult to comprehend for a 10-year-old.' But it sparked off his interest in exploring space from the ground.

Computer systems analyst Nick joined his local astronomical society in Sussex, going on to serve as its secretary. Now he is proud of the small observatory he has built in his Sussex back yard, where he and his wife, Linda, make their observations of the night skies. Linda, whom he met through the society, also joins the eclipse expeditions. While he focuses on close-ups of the Sun, her photographs are more panoramic, capturing the landscapes and earthly effects of an eclipse. But even with this team effort, Nick says there is no substitute for the real thing, experience of the moment as a human being. 'There is not a single photograph that can cover what you see with the naked eye,' he says. 'Anyone who hasn't seen an eclipse can't look at a photo and know what it is really like.'

The awe-inspiring moment of totality pictured from Guadeloupe in 1998. (Wendy Maunder)

Nick Quinn

Nick Quinn

Totality, using a longer exposure, captures more of the Sun's corona. (Java 1983)

Totality in Java, 1983.

It is the magnitude of experiencing an eclipse first-hand which Nick, like Francisco, hopes will captivate the British public when the 1999 event sweeps across our shores. He describes his office colleagues as a 'bit sceptical' of his hobby and his choice of strange holiday spots. 'They find it hard to understand anyone who spends days travelling across the globe to catch an event of a minute or so,' he says.

'They think it is a slightly eccentric thing to do – or maybe a very eccentric thing to do. But, for the people who wonder what all the fuss is about, I'll be interested to know what they think if they see the total eclipse of 1999 from the path of totality 'It's so hard to talk about the mightiness of this solar phenomenon to those who have never experienced it.'

Chapter Six

The Unexplainable

From the sky a great king of terror will come

Eclipses have played a role in shaping many of humankind's early superstitions and even today they are still believed by some to mark changes in the fate of the world.

There are those who believe that the total eclipse of 1999 heralds the open, completely public landing of extra terrestrials and the first meeting between the human race and an alien species in the months that follow. Throughout history there have been hundreds of thousands of reports of unidentified flying objects (UFOs) in the sky, both at night and in broad daylight. The majority of these sightings are in the USA, where there are many keen 'UFO watchers', but in all corners of the globe people have told of strange objects in the heavens – and a great many of them come during the time of an eclipse.

Some people even claim there is a link between UFO phenomenona and the prophecies of Nostradamus, who foretold of a great and terrifying king coming from the sky in 1999. Others see the eclipse and the millennium as heralding hunger and drought, or a new war across Europe. However far-fetched these claims might seem, UFO theories are gaining ground among many people. Indeed, over the last few decades, the number of sightings has grown dramatically, in particular around the time of solar and lunar eclipses and especially as we approach the millennium.

Sightings during eclipses are not confined to recent history and the age of cameras, powerful telescopes, instant communication and electronic tracking equipment. Reports of moving objects in the sky and strange lights appearing in front of the dark disc of the Moon have been recorded almost since records began. One of the earliest accounts is from AD 775, when a monk, living in what is now England, described seeing a bright light following the

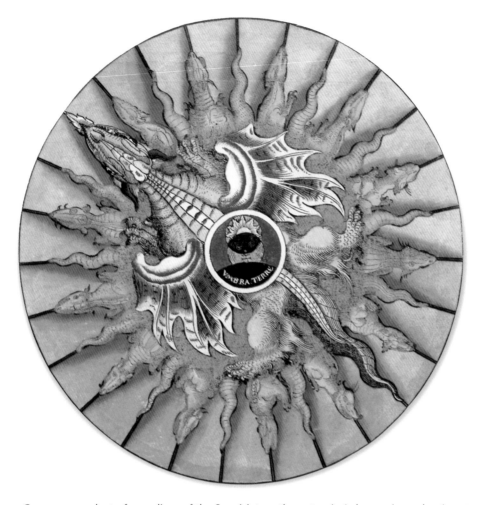

German woodcut of an eclipse of the Sun (sixteenth century). A dragon is predominent.

passage of the Moon. It accelerated past the lunar surface and then slowed down just moments before totality. In 1675, when North America still consisted of separate colonies, British soldiers reported seeing a strange, rectangular-shaped object in front of the Moon during a lunar eclipse, and in 1700 French astronomer M. Messier saw 'moving glows around the middle of the disc' as the earth blocked out the sun's light over the lunar surface. In the then colony of Australia, during a lunar eclipse in 1783, astronomer William Herschel reported seeing 'luminous points of light on or in front of the moon'. He saw the bright spots again in another eclipse 4 years later and, in 1790, wrote that he had seen '150 luminous points of red light on or near the moon'.

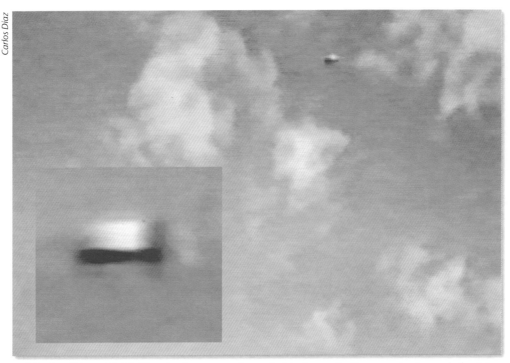

Carlos Diaz

Mexico, 8 June 1991. The single, saucer-shaped object seen by thousands of people for 13 minutes after the total eclipse. Inset: Close up detail.

Meanwhile, a crowd of townspeople in central France witnessed what was described as 'a most puzzling celestial event'. A formation of strange objects was reported to have been seen moving in a straight line. The objects were equally spaced and remained in line when they made turns. 'Their movements showed a military precision,' it was claimed. A formation of jet fighters perhaps? Not in 1820. In those days, planes and air travel were beyond most people's imagination.

During the nineteenth century, with continued advances in our knowledge of the universe and the building of specialized observatories housing more powerful telescopes and more accurate records, the sky was under surveillance more than ever before. Not surprisingly, the number of reported sightings increased, sightings which were not as easily dismissed as before, since many of the observers were scientists.

While watching the total solar eclipse of May 1836, Professor Auber, in Havana, Cuba, said he saw a 'considerable number of luminous bodies which appeared to move out from the sun in diverse directions.' And, during a lunar eclipse 12 years later, a Señor Rankin in Spain reported seeing 'points of light obscuring the Moon'. Studying a solar

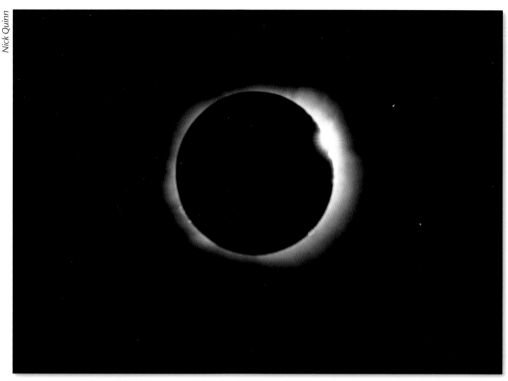

Total eclipse, Mexico, June 1991. A 'diamond ring' formation.

eclipse through a telescope in the state of Iowa, Professor Zentmayer wrote that he had 'observed some bright objects crossing from one cusp of the solar crescent to the other, each one taking two seconds to pass and all of them moving in straight lines, nearly parallel, in the same direction.'

With each successive eclipse came a flurry of sightings: bright lights speeding along in formation, huge balls of fire leaping from cloud to cloud and hurtling across the dark face of the eclipse, a single ray of light playing on the moon's surface.

On 29 July 1878, two astronomers, Lewis Swift (Director of the Warner Astronomical Observatory) and James C Watson (Director of the University of Michigan Astronomical Observatory) were studying a total eclipse of the Sun in Colorado through a telescope. They saw 'two, red disc-like objects cavorting some 32,180 km above the earth'. Another astronomer, who claimed to see the objects through a different telescope, estimated them to be at least 0.8 km in diameter.

While observing the total solar eclipse of 19 August 1887, two French astronomers, named Codde and Payan recorded 'an unknown body that appeared on the Sun's limb for 20 or 30 seconds after the eclipse.

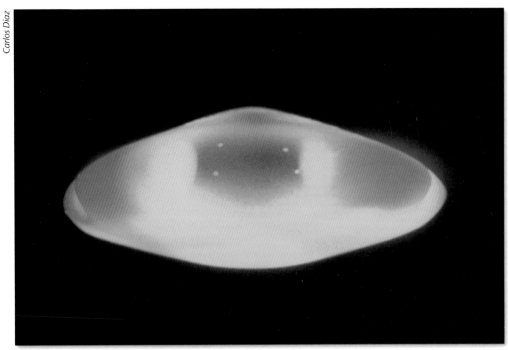

Carlos Díaz

Twelve days after the total eclipse of the Sun over Mexico in June 1991, this picture of a UFO was taken as sightings increased.

The body was round, with an apparent diameter about one-tenth that of the Sun.'

As the twentieth century dawned, humankind shakily took to the skies. New advances in camera technology and print processing meant that, for the first time, the images of strange flying objects could be captured not just in words but on photographic film – thus starting the century-long 'Are they real or aren't they?' wrangle. An early example of this occurred when, after recording a total solar eclipse in North Carolina in May 1900, astronomers Professor Pickering from Harvard and Mr Abbot of the Smithsonian Institute could not decide whether certain images on the photographic plate were due to unknown planets or defects in the film.

In 1954, potential UFO images which could not be dismissed simply as defective film, were photographed in Scandinavia. Two shiny, white rotating discs appeared on 10 seconds of a 16mm colour film shot from a plane. They were seen for about half a minute by 50 people in 3 separate aircraft, flying near Lifjell, Denmark. The witnesses included scientists and journalists taking part in an expedition to observe and film the total eclipse of the sun on 30 June of that year.

This marked the beginning of the global

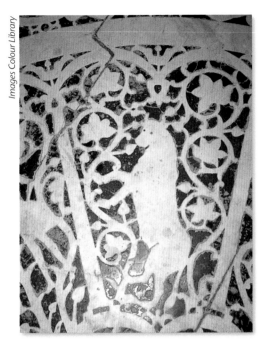

The image of Taurus from the marble zodiac at San Miniato al Monte, Florence, home of Galileo. Designed around 1207.

interest, the growth of a human fascination with stories of strange, alien objects of all shapes and sizes, flying in the skies above our planet – and without asking our permission first. It was around this time that the term 'flying saucer' came into being. Unidentified flying objects (UFOs), a term not widely adopted until the 1970s, is non-committal and seems to suit the unwillingness of governments and military establishments the world over to speculate on the subject.

All the sightings reported in the last 2,000 years were a drop in the ocean compared with the events which were to unfold in Mexico City.

Mexico City, with its 18 million inhabitants, is the most densely populated city on Earth. It was the optimum location for observing the total eclipse of the Sun on 8 June 1991. Tourists, astronomers and scientists all flocked in, eager to witness the solar phenomenon. Many came hoping that they would see the strange objects that seemed to have attended total eclipses before and they were not disappointed.

There was an electric atmosphere in the city that day. People waited eagerly. As early as noon, 1 hour before totality, it began getting darker, and just before the spectacular event, a wall of darkness – the Moon's shadow – could be seen moving towards the city. People in the street stood in silence and traffic slowed to a halt. It was 1pm.

But people had been staring for some time at something else. As the Moon began its passage in front of the Sun, a single saucer-shaped object appeared. It stopped between two large buildings in the city centre. The totality of the eclipse lasted about 7 minutes and, during the entire time and for a further 13 minutes afterwards, the object remained in place, silently, over the world's most populated city, during an event that had all eyes turned skywards.

The sighting was unique and witnessed by hundreds of thousands of people. The object was also photographed and filmed by many independent sources, from several different vantage points. Some shots clearly seemed to show a metallic, disc-shaped object in the sky.

del'Eclipfe, qui fera le 16. Septembre 1559. laquel-
le fera fa maligne extenfion inclufiuement, iuf-
ques à l'an 1560. diligemmēt obferuées par mai-
ftre Michel Noftradamus, docteūr en medecine
de Salon de Craux en Prouence. Auec vne fom-
maire refponce à fes detracteurs.

MDE
NOSTRE
DAME.

*Painting of the sixteenth-century astrologer Nostradamus in his studio with a decorative
border depicting the 12 images of the zodiac signs.*

Images Colour Library

Late fifteenth-century sculpture of Leo, on the Fitzjames Arch, Merton College, Oxford.

Images Colour Library

Fifteenth-century sculpture of Virgo in medieval dress on the Fitzjames Arch, Merton College, Oxford.

A week later, Jaime Maussan, a Mexican television journalist who hosted a TV show called *Sixty Minutes*, ran a special edition on the UFO sighting. He broadcast a video of the eclipse which included an image of a silvery disc-shaped object motionless above the city centre skyline. Maussan called on people who had similar recordings of the strange object to get in touch with him and, if possible, bring their tapes to the television station to be analysed. He received over 40,000 calls. The best quality recordings were studied on computer and showed the same digitalized image – a silver disc, domed, showing a red glow.

The Mexican 'flap', as it became known, caused much excitement among ufologists around the world and the sighting is perhaps the most heavily documented and studied incident of its kind ever. It also

*The Turkish astrological sign of Leo,
sixteenth century.*

that in the months following the August 1999 total solar eclipse, aliens will land and make open contact. Some say it is written in the stars.

The Babylonian priest Berosus wrote about creatures which were half-man and half-fish, and the Dogon, an African tribe in Mali, claim that their ancestors were taught by fish-gods who came from Sirius, the Dog Star. They called these aliens the Nommo.

Predictions and prophecies for human fortunes have attended eclipses for many centuries, extending right back to ancient and almost forgotten civilisations. Today's practical, technological world pays little attention to the pronouncements of soothsayers, whether or not they are connected to eclipses. Nevertheless, such predictions concerning the future of humankind are in abundance in a great many religions and many of the ancient practices are resurfacing within the New Age culture.

It was in the time before the industrial and technological revolutions, when people were very superstitious and lived in fear of making God angry, that the seers and fortune tellers flourished.

The New Age movement looks to prophecies made during this time as much as it does to new material from this century. Some beliefs come from sources as ancient as the Hopi and Mayan cultures. According to New Age observers, the Hopi Prophecy says that there will be serious change in the Earth and its inhabitants at the end of the twentieth century.

started a massive wave of UFO sightings in Mexico still going on today. Maussan became a committed UFO hunter and amassed evidence supporting more than 5,000 claimed sightings countrywide.

Today, following the strange events recorded in different parts of the world, many UFO enthusiasts and researchers are even more adamant that we are being visited and now watched quite openly by beings from outer space. Many proclaim

With hue like that when some great painter dips
His pencil in the gloom of earthquake and eclipse.
The Revolt of Islam

Percy Shelley

Take a pair of sparkling eyes,
Hidden, ever and anon,
In a merciful eclipse.
The Gondoliers

W.S. Gilbert

Mayan prophecies warn that by the year 2012, the Earth will come into synchronization with the universe. At this time the age of materialism, our preoccupation with property and owning things, must end. The prophecies predict we can then return to nature to save ourselves and the planet, our biosphere.

The Mexico eclipse is said to have fulfilled a Mayan prophecy, a solar total eclipse and mass UFO sightings. The Mayans, in AD 755, left a prophecy in their calendar that there would be a total solar eclipse on 11 July 1991. It said that this eclipse would be the harbinger of two things: 'Cosmic awareness through encounters with masters of the stars' and 'earth changes'.

The person with the most to say about future events, however, was Nostradamus, perhaps the most famous prophet of all and doyen of today's seers and soothsayers. They believe that his twentieth-century predictions have already come true – the 1978 Iran revolution, with the emergence of the Ayatollah and fall of the Shah of Persia; the first Gulf War between Iran and Iraq; the rise of Saddam Hussein; the invasion of Kuwait and the civil war in Yugoslavia. Now they wait for more.

Michel de Nostradame was born on 14 December 1503 at St Remy in Provence, France. The first son of Jewish parents, he was forced by the Inquisition to convert to Catholicism. Both his grandfathers taught him a wide range of subjects, and, by the time he began his formal education in Avignon, where he studied philosophy, grammar and rhetoric, he was already well versed in classical literature, history, astrology and herbal and folk medicine.

Graduating from the medical school at the University of Montpellier, Nostradamus

Sunbites

- The first solar research using a modern rocket was made in 1946, when a captured and converted German V2 rocket was launched from White Sands, New Mexico. It reached 55 km and recorded the solar spectrum.

- The first detailed solar observations from a manned spacecraft were made from the US station Skylab launched on 14 May 1973. Previously a little solar work had been carried out by Apollo astronauts who collected solar dust particles from the lunar surface.

began a private practice where he was renowned for treating plague victims with his potions of powdered rose petals. In 1534, he married and soon had a family, continuing his profession as a doctor even though by now he was gaining fame as a seer of the future.

The bubonic plague, also known as the Black Death, was still rampant in Europe and, tragically, it claimed the lives of his wife and two children. Distraught and pursued by the Inquisition because it was claimed by some that he was a sorcerer, Nostradamus packed his bags and, for the next six years, travelled throughout Italy and France. It was during this time that he began his work as a prophet.

Ten years after the death of his family, Nostradamus settled in Salon, where he remarried and had six children, turning the upper floor of his house into a private study. Here, he installed his special equipment, including an astrolabe, magic mirrors, divining rods, a brass bowl and a tripod.

At night he would sit before the tripod upon which simmered a brass bowl that had been filled with water and pungent herbs. He would then let thoughts sweep freely over him. For several years, Nostradamus struggled with the dilemma of whether or not he should make his prophecies public knowledge.

Then, in 1550, he published his first almanac – twelve four-line poems called 'quatrains'. Each quatrain gave a general prophecy for the coming year. The acclaim that he received encouraged him to continue his work and he produced an almanac every year for the rest of his life. But his most famous works, *The Centuries*, some of which refer to our own time, were started in 1554.

In *The Centuries*, which have remained in

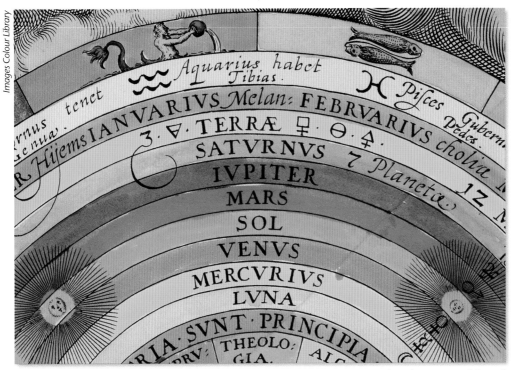

Zodiac signs of Aquarius and Pisces from an alchemical diagram printed in the 1677 edition of Museaeum Hereticum.

print for over 400 years, Nostradamus made many predictions about the fortunes of humanity in the later centuries of the millennium. Some, claim his supporters, appear to have come true, such as the rise of Adolf Hitler and the Challenger Space Shuttle disaster. Others appear to be far away from fact, and many are simply bewildering. They are written in rhyme form and contain some deeply obscure phraseology.

How he is interpreted is clearly a matter for the individual interpreter. A typical prediction is as follows:

The blood of the innocent will be a terror at London
Burned by thunderbolts, of twenty-three, the sixes,
The senile lady will lose her high position
Many more of the same sect will be slain.

The quatrain was written in 1555 and some followers believe it foretold the Great Fire of London in 1666 – the senile lady being St Paul's Cathedral, which was burned to the ground. Others say it refers to the burning of Protestant martyrs by a near-insane Queen Mary in England. The martyrs were sent to the stake in batches of 6,

Images Colour Library

The Men On The Moon

Eugene Andrew Cernan

- **born:** 14 March 1934, Chicago, Illinois
- **education:** B.S. in electrical engineering, Purdue University, 1956; M.S. in aeronautical engineering, Naval Postgraduate School, Monterey, California, 1964
- **spaceflights:** Gemini 9, Apollo 10, Apollo 17
- **subsequent career:** President, Cernan Corporations, aerospace management and marketing consulting

Harrison Hagan Schmitt

- **born:** 3 July 1935, Santa Rita, New Mexico
- **education:** BS in geology, California Institute of Technology, 1957; attended the University of Oslo, Norway, under a Fulbright Fellowship, 1957–58; PhD in geology, Harvard University, 1964
- **spaceflights:** Apollo 17
- **subsequent career:** a former US Senator; now science, technology and public policy consultant

sometimes with barrels of gunpowder (the thunderbolts referred to) between their legs for a more merciful end. But the number 23 is a mystery.

Of those prophecies remaining, most are supposed to happen within the next twenty years and, although many are too obscure to worry about, one of them is perhaps a little too close for comfort. In the months surrounding the 1999 eclipse, Nostradamus said that 'from the sky a great king of terror will come'. Some say that this means a giant meteor or comet will hit the Earth, causing catastrophic climate changes, earthquakes and tidal waves.

Scientists and astronomers have already identified the large comet, named FX11, heading in our direction. It is still a long way off and will not be close for another twenty years or so, they say. But exactly how close it will come, they don't yet know.

Most of Nostradamus's predictions are of doom and gloom. Fortunately, however, not everybody believes in their truth or accuracy. One thing we should not forget is that predictions, however accurate, do not have to come true. As individuals, we have free will and the ability to take action to change our own destiny.

The solar eclipse of 1999 and the new millennium are not just the territory of New Age predictions and UFOs, however. Sightings of ghosts, discoveries of holy relics, miracles and weeping statues are also on the increase. Even a holy vegetable – the aubergine – has featured in several cases. Apparently, when cut open, its seeds are said

Images Colour Library

The medieval zodiac sign of Pisces.

Moonbites

- The last man on the Moon was Eugene Cernan in Apollo 17. When he re-entered his lunar module it was the end of an era.
- There are about 3,000 moonquakes every year, centred mainly 600–800 km below the lunar surface.
- On 25 June 1178, a monk in Canterbury recorded that 'a flaming torch sprang up, spewing out fire, hot coals and sparks' from the edge of the Moon. It was probably a huge meteorite.

to have spelt out the Arabic word for Allah.

Some religious leaders are already using the language of the apocalypse – of the Book of Revelations, in which St John the Divine predicted a thousand-year reign by Christ and his saints on Earth, before Satan was set loose upon the world for the Last Judgement. Pope John Paul II has already told a Polish congregation: 'We are facing the final confrontation between the Church and the Anti-Church, between the Gospel and the Anti-Gospel.'

As bookmakers record an increase in bets on the date of the end of the world, will Cornwall and Devon follow in the footsteps of Mexico in the years after the August eclipse and be the site of many more reported sightings of UFOs? And will we ever learn the truth behind these strange occurrences?

Only time will tell.

Safety First
Viewing the Eclipse

Never stare at the Sun with the naked eye, Galileo went blind doing it. Our local star's ultraviolet and infra-red radiation will damage your eyesight, even if you do not feel any pain or irritation. Even when the Sun looks dull or is shrouded in mist that covers our planet, it is a dangerous and phenomenally powerful nuclear reactor, a massive fireball of hot gases and radiation that will fry your retina. The effect may not be immediate. Sometimes it can take years to develop into chronic eyesight problems.

Risking your eyesight to watch an eclipse of the Sun without taking adequate precautions, no matter how wondrous the event, is simply not worth it. Your eyes are too important. You cannot buy new ones.

The only time it is safe to look directly at the Sun with the naked eye is during totality – but this is risky, unless you really know what you are doing, because the passage from totality, when the dark disc of the Moon completely blocks out the Sun, to the partial phase, happens rapidly. Even looking at the Sun with an optical aid is potentially dangerous unless you really know what you are doing. If you do not know, DO NOT DO IT!

There are ways of viewing the partial phases of the eclipse safely, however. One method is known as 'pinhole projection'. All you need are two pieces of stiff cardboard or tin plate. Punch a clean pinhole in one of the pieces. Then, standing with your back to the Sun, hold the two cards about a metre apart so that sunlight passes through the hole in the first piece and is projected onto the second. The Sun's inverted image will be seen around 1 cm in diameter. Make sure that the pinhole you have punched is not too wide otherwise there will be no image. You can make the image bigger or smaller by adjusting the distance between the two cards. Remember: keep your back to the Sun at all times and DO NOT try to look at it through the pinhole with the naked eye.

Most people experiencing an eclipse will be advised to use solar viewers made of aluminized mylar. Mylar is a very tough plastic film, a few microns thick, that is amazingly strong. Solar filters are made by coating the film with a thin layer of aluminium metal. Mylar coated on both sides is the safest, but single-sided is better for photography because of the shorter exposure times allowed. However, taking

pictures of an eclipse is a specialized task and should not be attempted without proper guidance, equipment and training.[1]

If you plan to use mylar viewers they must be checked for flaws which could allow direct sunlight to reach your eyes. The best type of filter to use is one with two layers of aluminized mylar film, which cuts down the possibility of a weakness in one of the aluminium coatings resulting in damage

to your eyesight. If there is even so much as a pinhole through the film, damaging radiation will shoot into your retina. So check and double-check.

And, remember, sunglasses and binoculars of any type must not be used for looking at the Sun.

---

[1] A useful publication for guidance on eclipse photography is *The Sun in Eclipse* by Michael Maunder and Patrick Moore, published by Springer-Verlag.

ACKNOWLEDGMENTS

We would like to thank:

Steve Bell, of HM Nautical Almanac Office for his help and advice and for his publication, *The RGO Guide to the 1999 Total Eclipse of the Sun*, published by HMSO, which was an invaluable reference guide.

Michael Maunder, BSc, Cchem, FRSC, FRAS, MRI, whose pictures and help with this publication have also been invaluable. Also his book, *The Sun in Eclipse,* published by Springer-Verlag, London Ltd., and written in conjunction with Patrick Moore, CBE, FRAS, has been a valuable source of information.

Photographer Nick Quinn, whose pictures and stories have been so helpful.

Francisco Diego, Senior Research Fellow of the Department of Physics and Astronomy, University College, London, who has been a great help in the gathering of pictures and information.

Professor Richard Stephenson, Department of Physics, Durham University, whose research into the history of eclipses has enabled us to formulate a brief outline of the solar phenomenon since early records. A special mention, too, for his authoritative book, *Historical Eclipses and Earth's Rotation*, Cambridge University Press, 1997.

The British Newspaper Library, Colindale, North London for endlessly digging out microfilm on reports of Total Eclipses. The Royal Astronomical Society; the British Astronomical Society; the British Library; the National Maritime Museum; the British Association for the Advancement of Science; Cambridge University Press; the Bridgeman Art Library and the Press Association of London.

NASA in Washington and Houston ,Texas, for ceaseless friendliness and help in gathering information and pictures on the Solar System and the Apollo Moon programme. Special thanks to Fred Espenak, Stephen Edberg, Tom Farmer and Neil Armstrong.

Researchers Hilary Krieger in Boston and Michael Dean in Belfast.

Philip Mantle, President of the British UFO Research Association and his colleague and friend, Michael Hesemann, in Dusseldorf.

Michael Maunder

BIBLIOGRAPHY

Special reference material from *The Secret Language of the Stars and Planets* by Geoffrey Cornelius and Paul Devereux, Duncan Baird Publishers; *The Earth's Moon* by Isaac Asimov, published by Gareth Stevens of London; *Chronology of Eclipses and Comets, AD 1-1000* by D Justin Schove, published by The Boydell Press of Suffolk; *Reaching for the Stars* by Peter Bond, published by Cassell, London; *Heavens Unearthed* by Matt Kane, published by Golden Egg Books, America; *Greek Mythology*, an encyclopedia by Richard Stoneman, published by The Aquarian Press, London; *Indian Mythology* by Jan Knappert, also published by Aquarian; *In the Beginning: The Birth of the Living Universe*, published by Penguin Books; *Greece and Rome: Myths and Legends* by H.A. Guerber, published by Senate, London; *The Universe* by James Muirden, published by Kingfisher Books; London; *The Universe for Beginners* by Felix Pirani and Christine Roche, published by Icon Books, Cambridge; *Stars and Planets* by James Muirden, also published by Kingfisher; *Manned Spaceflight*

Log by Tim Furniss, published by Jane's of London; *Stonehenge* by RJC Atkinson, published by Penguin Books; *The Moon* by Lesley Sims, published by Watts Books of London; *Our Star The Sun*, edited by Maria O'Neill, published by Belitha Press Ltd, London; *Astronomy* by Kristen Lippincott, published by Dorling Kindersley; *The Cambridge Illustrated History of Astronomy*, edited by Michael Hoskin, Cambridge University Press; *World Mythology, The Illustrated Guide*, edited by Roy Willis, published by Duncan Baird, London; *Patrick Moore's History of Astronomy*, published by Macdonald and Co, London; *New Bible Dictionary*, 2nd edition, edited by JD Douglas, published by Intervarsity Press, Leicester; *The HarperCollins Bible Dictionary*, edited by. Paul J. Achtemeier, published by HarperCollins Publishers, New York; *Dictionary of Gods and Goddesses, Devils and Demons*, edited by Manfred Lurker, published by Routledge & Kegan Paul, London; *The Illustrated Encyclopedia of Myths and Legends*, edited by Arthur Cotterell, published by Cassell Publishers Ltd., London; *The Goddesses and Gods of Old Europe* by Marija Gimbutas, published by Thames & Hudson, London; *Eclipse of the Sun* by Janet McCrickard, published by Gothic Image Publications, Somerset; *Moon Lore* by Timothy Harley, published by Swan Sonnenschein, Le Bas & Lowrey; *Earth and Sky: Visions of the Cosmos in Native American Folklore*, edited by Claire R Farrer and Ray A Williamson, published by University of New Mexico Press, Albuquerque; *The Secret of the Incans: Myth, Astronomy and the War Against Time* by William Sullivan, published by Crown Publishers, Inc., New York; *Myth, Legend and Custom in the Old Testament, Volume 11* by Theodor H Gaster, published by Harper & Row, Publishers, New York; *Legends and Superstitions of the Sea and Sailors* by Flethcher S. Bassett, published by Singing Tree Press, Detroit; *Dictionary of Jesus and the Gospels*, edited by Joel B Green and Scot McKnight, published by Intervarsity Press, Leicester; *A Chronicle of Pre-Telescopic Astronomy* by Barry Hetherington, published by John Wiley & Sons, West Sussex; *Myths of the Middle Ages* by Sabine Baring-Gould, published by Cassell Plc., London; *The Faber Book of Northern Legends*, edited by Ed. Kevin Crossley-Holland, published by Faber & Faber; *The Holy Bible*, published by Burns & Oates, London; *Beyond Stonehenge* by Gerald Hawkins, published by Hutchinson & Co Ltd., London; *The International Standard Bible Dictionary, Volume One*, edited by Geoffrey W Bromiley, published by William B Eerdmans Publishing Company, Michigan.